Solved Problems in Geostatistics

Solved Problems in Geostatistics

Oy Leuangthong
University of Alberta
Department of Civil & Environmental Engineering
Edmonton, Alberta, Canada

K. Daniel Khan
Chevron Energy Technology Company
Houston, TX

Clayton V. Deutsch
University of Alberta
School of Mining and Petroleum Engineering
Edmonton, Alberta, Canada

A JOHN WILEY & SONS, INC., PUBLICATION

Published by John Wiley & Sons, Inc., Hoboken, New Jersey.
Published simultaneously in Canada.

Library of Congress Cataloging-in-Publication Data:

Leuangthong, Oy, 1974–
Solved problems in geostatistics / Oy Leuangthong, Daniel Khan, Clayton V. Deutsch.
p. cm.
Includes bibliographical references and index.
ISBN 978-0-470-17792-1 (pbk.)
1. Geology—Statistical methods—Problems, exercises, etc. 2. Geology—Data processing—Problems, exercises, etc. I. Khan, K. Daniel, 1972– II. Deutsch, Clayton V. III. Title.
QE33.2.S82L48 2008
550.72--dc22 2008007625

CONTENTS

PREFACE AND ACKNOWLEDGMENTS

Reliable application of geostatistics for modeling regionalized variables requires knowledge about geostatistics and a lot of practice. There are many styles of learning, but a common thread is that people learn by working through problems. This book provides a set of problems that outline the central themes of geostatistics with worked solutions. Our hope is that this collection of problems will help students gain familiarity with the overarching principles, the common applications, and some of the intricate details of this rich subject.

The authors wish to acknowledge the teachers and researchers who have contributed to the growth and evolution of geostatistics. Some of these problems are derivatives of André Journel's from back in the 1980s.

Oy would like to thank Shawn for always being so supportive in this and all other endeavours. She would also like to thank her son Brayden who, although too young to realize, provides the motivation to be efficient. Together they help provide a happy and balanced life.

Dan would like to acknowledge the selfless support of Joanna throughout his early career endeavors; the years add up quickly. Thanks also to his children Sofia, Aaron, and Sarah, who endured his absence during working hours.

Clayton would like to acknowledge Pauline for her enduring support for more than 20 years of geostatistical adventure. Jared read through most chapters and had some good input. Matthew and Rebecca complete the family. Clayton thanks them all.

CHAPTER 1

Introduction

Geostatistics has become increasingly popular for numerical modeling and uncertainty assessment in the earth sciences. The beginnings of geostatistics can be traced to innovation in the mining industry in the late 1950s and early 1960s. Application of geostatistics has expanded into other industries including reservoir characterization, meteorology, environmental, hydrogeology, and agriculture. The common features of problems addressed by geostatistics include variables regionalized in space, sparse data, and nonrandom patterns of heterogeneity.

The number of geostatistical theoreticians and practitioners is growing. A reasonable number of courses, books, and papers are available that document the techniques and principles of geostatistics. These include Armstrong (1998), Chilès and Delfiner (1999), Christakos (1992), Clark and Harper (2000), Cressie (1991), David (1977), Davis (1986), Deutsch (2002), Goovaerts (1997), Hohn (1988), Isaaks and Srivastava (1989), Journel (1989), Journel and Huijbregts (1978), Mallet (2002), Matern (1960), Matheron (1969, 1971), Yarus (1995), and Wackernagel (2003). Despite the resources available to the student of geostatistics, the discipline of geostatistics is relatively small and fast growing. There is no comprehensive textbook that explains theory, practice, and provides solved problems to guide the student through the learning process. This book provides a tour through important problems in geostatistics.

This book is aimed at a *student* of geostatistics. We are all students; some are just more formally recognized as such: senior undergraduate students and graduate students. Senior practicing geostatisticians interested in understanding the theoretical principles would be less formally recognized students. Young professionals who recently graduated in the geosciences or earth science related engineering are a special target of this book. Unfortunately, many universities have no requirement for classes in numerical geological modeling with

Solved Problems in Geostatistics. By O. Leuangthong, K.D. Khan, and C.V. Deutsch

geostatistics. The young professional faced with the task of geostatistical modeling without a formal class in the subject, needs a few key references. This book is one of those references.

Most people benefit from seeing how problems are solved. Looking through the problems provides an overview of the discipline. Reviewing the solution methodology provides an overview of the mathematical and numerical tools used in the discipline. Our goal is to present a collection of problems, some theoretical, some practical, that define the key subject areas of geostatistics. These problems point to specific textbook references and specialized papers that, collectively, more completely define the discipline.

Most of these problems have been used in some variant of undergraduate courses, graduate courses, or industry short courses. The authors attended and/or teach at the University of Alberta and Stanford University. The problems lean toward assignments encountered at these two schools and are considered important by the authors. The problems have been modified to focus on key principles and to target a technologically evolving audience. There are many more problems in geostatistics than we could possibly cover in this book.

There are three different types of problems: (1) analytical problems that can be solved by hand, (2) numerical problems that can be solved using any generic spreadsheet type of software, and (3) practical problems that often require some specialized geostatistical software. In the latter case, the intent is not to focus on any particular public domain or commercial geostatistical package; the reader should be able to use any geostatistical software.

For each problem, the objective is given at the beginning to highlight the key learning that is intended. This is followed by the background and assumptions required to proceed with the exercise. The problem is then described with any required figures and/or graphs. A solution plan is then presented that should walk the reader through to the correct solution. This plan is only provided to facilitate the learning process; readers should attempt to solve the problem without this solution plan if possible. In most cases, a partial solution is then presented to permit the reader to check his/her individual solution. Finally, some closing remarks are given with respect to some key ideas that are related, but may not have been required to perform the exercise.

1.1 PLAN OF THIS BOOK

There are nine core chapters. Each chapter is aimed at a particular topic and consists of three problems. We expect students to take from 1 to 20 hours to work through their own solution to each problem.

Chapter 2 addresses some basic probability concepts. Although the use of parametric distributions is diminishing in modern practice, the ability to calculate moments and quantiles of analytical distributions provides valuable insight into the probabilistic paradigm underlying geostatistics. The first problem calls on basic probability theory. Weighted combinations of data are used commonly in geostatistics. The second one relates to calculating the variance of a linear combination. Finally, the transfer of data and statistics between standardized and nonstandardized units is considered in the last problem.

Chapter 3 focuses on the need for representative statistics. Sites are not sampled equally and considering data as equally representative independent samples from an underlying population is unrealistic. The first problem is a basic one that shows how unequal weights are assigned to data. The second one is a theoretical problem related to the practical notion that there are parts of the distribution that may not have been sampled. In this case, secondary data can be used to establish a representative distribution. The final problem relates to a comparison of practical tools for determining representative statistics.

Chapter 4 reveals the fundamental tool of Monte Carlo Simulation (MCS). The first exercise uses MCS to demonstrate the Central Limit Theorem, that is, the sum of identical equally distributed random variables tends to a Gaussian distribution regardless of the starting distribution. The bootstrap and spatial bootstrap are introduced in the second problem as methods for assessing parameter uncertainty. The third problem is a practical one related to the transfer of uncertainty from input random variables to some nonlinear response variable of interest.

Chapter 5 presents the venerable variogram. The notion of geometric anisotropy is remarkably simple until faced with a practically difficult three-dimensional problem, and then the intricacies of dealing with anisotropy become far more challenging. The first problem explores the details of geometric anisotropy used throughout geostatistics. The second one requires the student to calculate a variogram by hand from a small dataset. Finally, the third problem calls for variogram modeling and the use of the variogram in understanding how variance decreases as scale increases.

Chapter 6 exposes kriging as an optimal estimator. Kriging was independently developed by many experts in many different areas of application. The first problem asks the student to derive the essential equations underlying the kriging estimator. Many variants of kriging are based on constraints to achieve unbiasedness under different models for the mean. The second one asks the student to derive the basis of ordinary, universal and external drift kriging. Kriging is a unique mathematical technique that accounts for proximity of the data to the unsampled value and redundancy between the data. The final problem explores some of the properties of the kriging estimator.

Chapter 7 focuses on Gaussian simulation and its unquestioned importance in modeling uncertainty in continuous variables. The first problem explores the bivariate Gaussian distribution. This setting is good for learning about the properties of the multivariate Gaussian distribution since it is difficult to visualize higher order distributions. The second problem brings out the importance of conditioning in geostatistics; that is, enforcing the reproduction of local data. The use of kriging to condition unconditional realizations is revealed. Finally, a more complete problem set is presented for simulation of practical-sized realizations.

Chapter 8 is devoted to problems of indicator geostatistics. The rich depths of categorical variable modeling are touched by indicators. The first problem relates to the theoretical link between indicator variograms and object sizes and shapes. The second one demonstrates the use of indicator variograms in the context of checking a multiGaussian assumption. The third problem asks for a hand calculation to perform indicator kriging for the construction of a conditional probability distribution of a categorical variable.

Chapter 9 explores the details of modeling more than one variable simultaneously. The first problem demonstrates the simultaneous fitting of direct and cross variograms with the only practical "full" model of coregionalization: the linear model of coregionalization. The second problem applies the well established paradigm of cosimulation in a multivariate Gaussian framework. In the last problem of this chapter the challenging problem of modeling multiple variables at different scales is introduced.

Chapter 10 touches on a few newer topics. The use of utility theory and loss functions combined with geostatistically-derived distributions of uncertainty is applied in the first problem. The second one is related to the use of probability combination schemes, such as permanence of ratios. These have gained in popularity and are applicable to modeling trends of categorical variables. Finally, the last exercise relates to multiple point geostatistics, which has become increasingly popular in recent years.

These 27 problems highlight geostatistical methods that are either well known or form the basis for techniques that have emerged as particularly promising for future research and application. Some of the problems require data or specific programs. These files and other supplementary material are available from the website **www.solvedproblems.com**. Corrections to the problems and solutions together with additional problems are also available from this website.

1.2 THE PREMISE OF GEOSTATISTICS

At the time of writing this book, the philosophical framework and toolset provided by geostatistics provides the best approach to predict spatially distributed variables. The challenge of spatial inference is overwhelming. Less than one-trillionth of most geological sites are sampled before we are asked to provide best estimates, quantify uncertainty, and assess the impact of geological variability on engineering design.

Matheron formalized the theory of geostatistics in the early 1960's (Matheron, 1971). Geostatistics was not developed as a theory in search of practical problems. On the contrary, development was driven by engineers and geologists faced with real problems. They were searching for a consistent set of numerical tools that would help them address real problems, such as ore reserve estimation, reservoir performance forecasting, and environmental site characterization. Reasons for seeking such comprehensive technology included (1) an increasing number of data to deal with, (2) a greater diversity of available data at different scales and levels of precision, (3) a need to address problems with consistent and reproducible methods, (4) a belief that improved numerical models should be possible by exploiting computational and mathematical developments in related scientific disciplines, and (5) a belief that more responsible decisions would be made with improved numerical models. These reasons explain the continued expansion of the theory and practice of geostatistics. Problems in mining, such as unbiased estimation of recoverable reserves, initially drove the development of geostatistics (Sichel, 1952; Krige, 1951). Problems in petroleum, such as realistic heterogeneity models for unbiased flow predictions, were dominant from the mid-1980s through the late-1990s. Geostatistics is extensively applied in these two areas and is increasingly applied to problems of spatial modeling and uncertainty in environmental studies, hydrogeology, and agriculture.

The uncertainty about an unsampled value is modeled through a probability distribution. These probability distributions are location dependent. The set of probability distributions over the domain of interest defines a random function (RF), which is the central aim of geostatistics. Inference of statistics related to a RF requires a choice of how to pool data together for common analysis. Furthermore, we may have to model large-scale trends because most regionalized variables exhibit large-scale variations. The decision of *stationarity* is this combination of (1) a choice of how data are pooled together, and (2) the location dependence of spatial statistics. The suitability of geostatistical modeling for its intended purpose requires a reasonable decision of stationarity. It is difficult to conceive of a problem set that would reveal just how important stationarity is to geostatistics.

Geostatistics is concerned with constructing high-resolution models of categorical variables, such as rock type or facies, and continuous variables, such as mineral grade, porosity, or contaminant concentration. It is necessary to have *hard* truth measurements at some volumetric scale. All other data types including remotely sensed data are called *soft* data and must be calibrated to the hard data. It is neither possible nor optimal to construct models at the resolution of the hard data. Models are generated at some intermediate geological scale and then upscaled for computationally intensive process performance. A common goal of geostatistics is to construct detailed numerical models of geological heterogeneity that simultaneously account for a wide range of relevant data of varying resolution, quality, and certainty, so much of geostatistics relates to data calibration and reconciling data types at different scales.

1.3 NOMENCLATURE

The following list consists of fairly standard nomenclature used in a wide variety of geostatistical literature.

C(**h**): Covariance between two random variables separated by vector **h**
$Circ_a$(**h**): Circular variogram model of parameter a for points separated by **h**
Cov{X,Y}: Covariance between X and Y
$D^2(v, V)$: Dispersion variance of samples of scale v within volumes V
E{•}: Expected value of •
f(z): Probability density function of a random variable Z
F(z): Cumulative distribution function of a random variable Z
γ(**h**): Semivariogram between two random variables separated by vector **h**
$\bar{\gamma}(v, V)$: Average semivariogram between volumes v and V
Γ(**h**): Standardized semivariogram (sill of 1.0)
G(y): Standard normal or Gaussian cumulative distribution function
h: Vector separation distance between two points
h: Scalar distance between two points
i(**u**;k): Indicator value for category k at location **u**
k: Index of a particular rock type or category
$L(z^*-z)$: Loss function for error z^*-z (estimate minus truth)
λ_α: Kriging weight applied to datum α
μ: Mean of a statistical population
p_k: Proportion of rock type or category k
Prob{•}: Probability (proportion over similar circumstances) of •
ρ: Correlation coefficient (standardized covariance) between two variables
σ^2: Variance
$\sigma_K{}^2$: Kriging variance associated with estimating a particular locaiton
Sph(**h**): Spherical semivariogram function for vector **h**
u: A location in space (1, 2, or 3D)
Var{•}: Variance of •, that is, the expected squared difference from the mean
w: Weight assigned to a data (often from declustering)
Z: Generic random variable
z: A particular outcome of the random variable Z
z_k: A particular threshold (the kth one) of the random variable Z
Z(**u**): Generic random variable at location **u**

Additional notation will be defined in the text as needed. Readers may refer to Olea (1991) for a more extensive and general glossary of geostatistical terminology.

CHAPTER 2

Getting Comfortable with Probabilities

The language of statistics and probability is universally used to present uncertainty. This chapter introduces some of the important and relevant terminology found in geostatistics. The first exercise establishes the link between the probability density function (pdf), the cumulative distribution function (cdf), and reviews the calculation of the mean and variance. The second exercise looks at the statistics of a linear combination of random variables (RVs), which is a recurring theme in geostatistical tools. Finally, the third exercise is aimed at gaining comfort with the units of a Gaussian variable, moving between standardized and nonstandardized units and also considering the probability interval as another dispersion statistic.

2.1 PARAMETRIC PROBABILITY DISTRIBUTIONS

Learning Objectives

This problem explores the statistical analysis of a simple parametric distribution. While real data rarely follow a parametric distribution, the principles of cumulative distributions and probability density functions are the same. Moreover, the mean and variance are two summary statistics with very clear,

Solved Problems in Geostatistics. By O. Leuangthong, K.D. Khan, and C.V. Deutsch

formal definitions that apply regardless of whether we consider a parametric or nonparametric distribution.

Parametric distributions are fully defined by a small number of parameters. A nonparametric distribution is defined directly from the available data by the proportions less than specified thresholds. Parametric distributions are mathematically congenial and used in many circumstances (Johnson and Kotz, 1970). The main objective of this problem is to gain a deeper understanding of the fundamental properties of probability distributions and important summary statistics. A secondary objective is to review some of the mathematics required in statistics and geostatistics.

Background and Assumptions

The central paradigm of statistics is to replace the unknown truth by a random variable, Z. Our uncertainty in Z is represented by a probability distribution. The cdf $F(z)$ is one representation of uncertainty for continuous variables. The cdf is defined as the probability that the RV is less than a threshold, for increasing thresholds:

$$F(z) = \text{Prob}\{Z \leq z\}, \qquad z \in [-\infty, \infty]$$

The cdf at a particular threshold z can be thought of as the proportion of times the random variable Z would be less than the threshold in similar circumstances. The cdf in any particular setting will be estimated by (1) a degree of belief based on some understanding of the problem, (2) proportions from data taken in similar circumstances, or (3) a mathematical model. A mathematical model is often fitted to the data to facilitate calculations and inference to unsampled locations.

The interpretation of the cdf as a proportion leads to some mathematical properties. These properties are directly related to the fact that proportions cannot be negative and they must sum to one in a closed set. The cdf is a non-decreasing function, there is no proportion less than minus infinity $F(-\infty) = 0.0$, and the random variable must be less than infinity $F(+\infty) = 1.0$.

The cdf is used directly in many statistical and geostatistical calculations. At times, however, it is convenient to think of the pdf, $f(z)$. The pdf is defined as the derivative of the cdf if it exists:

$$f(z) = F'(z) = \lim_{dz \to 0} \frac{F(z+dz) - F(z)}{dz}$$

Through the fundamental theorem of calculus, the integral of the pdf is the cdf. The properties of the cdf lead to the following properties of the pdf: (1) $f(z) \geq 0 \quad \forall z$, and (2) $\int_{-\infty}^{+\infty} f(z)dz = 1.0$. Given a valid pdf, we can calculate cumulative probabilities associated to certain outcomes as:

$$\begin{aligned} F(z_1) &= \int_{-\infty}^{z_1} f(z)dz \\ &= \text{Prob}\{Z \leq z_1\} \end{aligned} \tag{2.1}$$

These mathematical properties of the cdf and pdf are axiomatic given the interpretation of probabilities as proportions. The central problem in practice is to establish the distribution of uncertainty at an unsampled location.

Uncertainty in the case of a discrete variable is simpler. Consider a random variable Z that must take one of a discrete set of outcomes: s_k, $k = 1,...,K$. The uncertainty in Z is fully specified by the probability of each outcome:

$$p_k = \text{Prob}\{Z = s_k\}, \qquad k = 1,...,K$$

These probabilities are interpreted as the proportion of times the random variable Z would be the outcomes s_k, $k=1,...,K$ in similar circumstances. They must be nonnegative and sum to 1.0. In some cases, for mathematical convenience, the probabilities may be ordered and summed to form cumulative probabilities like a cdf.

Although the probability distribution of a random variable fully states what we know (and do not know) about the unknown truth, it is common to summarize the distribution with certain statistics. These statistics mostly apply to continuous variables; there is often no inherent numerical meaning to discrete variables. The mean and variance of the RV can be calculated by considering the expected value operator, $E\{\cdot\}$. The expected value or mean of a random variable is defined as:

$$E\{Z\} = m = \int_{-\infty}^{+\infty} z \cdot f(z)dz \tag{2.2}$$

The mean is also known as the first-order moment of the RV. For practical purposes, the mean can be thought of as the probability weighted average of the RV.

The variance is considered as the second-order moment and is defined as

$$\begin{aligned} Var\{Z\} &= \sigma^2 \\ &= E\{(Z-m)^2\} \\ &= E\{Z^2\} - m^2 \end{aligned} \tag{2.3}$$

where m is the mean as given by Equation (2.2). The variance is also known as a second-order moment of the RV. The mean is a summary statistic that gives the center of the distribution. The variance is a measure of dispersion from the mean. These are the two most widely used summary statistics for continuous variables.

Problem

For a simple triangular distribution, we can express the density function, the cumulative distribution, and its moments in terms of the simple geometrical properties of the density curve shown in Figure 2.1.

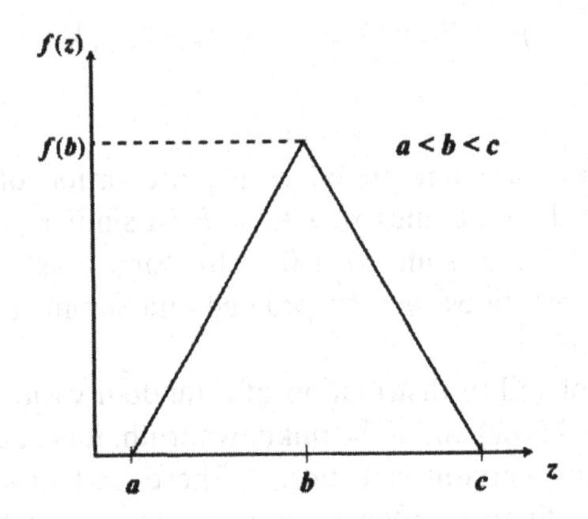

Figure 2.1 Triangular probability density function defined by three parameters: a, b, and c.

With the use of the functional form of the triangular pdf given in Equation (2.4), (1) write an expression for the cdf of Z; (2) express $f(b)$ in terms of the parameters a, b, and c; and (iii) write expressions for the first two moments of Z (i.e., the mean and variance).

$$f(z)=\begin{cases}0\,, & \forall\ z<a\\ \left(\dfrac{z-a}{b-a}\right)f(b) & a\le z\le b\\ \left(1-\dfrac{z-b}{c-b}\right)f(b) & b\le z\le c\\ 0\,, & \forall\ z>c\end{cases} \tag{2.4}$$

Solution Plan

(i) Determine the cdf based on the definition of the cumulative distribution of a RV as shown in Equation (2.1). This expression will evaluate the integral over all relevant intervals between $-\infty$ to $+\infty$.

(ii) A licit pdf has the property that its integral, computed between $-\infty$ to $+\infty$ equals 1.0, or equivalently, the area under the curve equals unity. The value of $f(b)$ must ensure that this property holds true for this pdf.

(iii) To calculate the mean, integrate Equation (2.2) for $E\{Z\}$ and evaluate over the limits $[a,b]$ and $[b,c]$. Similarly, to calculate the variance, integrate Equation (2.3) for $Var\{Z\}$ and evaluate over the limits $[a,b]$ and $[b,c]$.

Solution

(i) The definition of the cdf $F(z)$ is $F(z)=\int_{-\infty}^{z} f(z)\,dz$. Substitute the pdf expression $f(z)$, from Equation (2.4), into this formal definition of the cdf $F(z)$. This expression of the cdf should have two components evaluated between $[a,b]$ and $[b,c]$.

(ii) For $f(z)$ to be a licit pdf, $F(z)=\int_{-\infty}^{+\infty} f(z)\,dz=1$, that is, the area under the curve (or the pdf) must be equal to 1.0. Knowing this property yields the following solution:

$$f(b)=\frac{2}{c-a} \tag{2.5}$$

(iii) To evaluate the mean of the distribution, apply the definition of the expected value:

$$
\begin{aligned}
E\{Z\} &= \int_{-\infty}^{+\infty} z\, f(z)dz \\
&= \int_a^b z\frac{z-a}{b-a} f(b)dz + \int_b^c z\left(1-\frac{z-b}{c-b}\right) f(b)dz \\
&= \frac{f(b)}{b-a}\int_a^b (z^2 - az)dz + \frac{f(b)}{c-b}\int_b^c (cz - z^2)dz
\end{aligned}
\tag{2.6}
$$

Integrating the terms of Equation (2.6) between their respective limits, and substitution via Equation (2.5) gives

$$
E\{Z\} = m = \frac{a+b+c}{3} \tag{2.7}
$$

In a similar workflow, the variance of the distribution can be evaluated as:

$$
\begin{aligned}
Var\{Z\} &= E\{Z^2\} - m^2 \\
&= \frac{f(b)}{b-a}\int_a^b z^2(z-a)dz + f(b)\int_b^c z^2\left(1-\frac{z-b}{c-b}\right)dz - m^2 \\
&= \frac{f(b)}{b-a}\int_a^b (z^3 - az^2)dz + \frac{f(b)}{c-b}\int_b^c (cz^2 - z^3)dz - m^2
\end{aligned}
\tag{2.8}
$$

Integrating the terms of Equation (2.8) between their respective limits, and substitution via Equation (2.5) gives

$$
Var\{Z\} = \frac{1}{18}(a^2 + b^2 + c^2 - ab - bc - ac) \tag{2.9}
$$

Remarks

In the presence of sparse data, the triangular distribution is often used because it is a continuous pdf with a clearly defined lower and upper limit with a non-uniform density. These properties make it convenient to work with, particularly when the underlying distribution of a variable is not known other than some estimates of the range, $[a, c]$, around a most-likely estimate, b. A uniform distribution might be chosen to model uncertainty in a random variable where there is insufficient information available to suggest a best estimate.

This exercise is a classical introduction to the derivation of the first two moments of a distribution via elementary calculus. The mean and variance are two primary statistics that are integral to the modeling approaches applied in geostatistics. Specifically, stationarity of these two statistics have been a fundamental assumption in the development of geostatistical tools.

2.2 VARIANCE OF LINEAR COMBINATIONS

Learning Objectives

In many fields of application, including geostatistics, the use of linear combinations of random variables is fundamental to addressing the problem of statistical inference. The uncertainty associated with the linear combination is interesting in presence of correlation. The purpose of this exercise is to gain insight on uncertainty in the presence of correlation and independence through finding an expression for the variance of a weighted average of RVs. This exercise also provides insight into parameter uncertainty. It is common in geostatistics to deal with few data for inference; uncertainty in the mean of a distribution is important and may need to be transferred through subsequent calculations. This exercise demonstrates the uncertainty of the mean in the presence of correlated data.

Background and Assumptions

Consider n data, $\{z_i, i=1,...,n\}$ and the mean of these data, $\bar{z}$:

$$\bar{z} = \frac{1}{n}\sum_{i=1}^{n} z_i \tag{2.10}$$

in the case of equally weighted samples. An expression for the general weighted average estimator is

$$z^* = \sum_{i=1}^{n} \lambda_i z_i \tag{2.11}$$

that would be the case in the presence of declustering weights (see Chapter 3). The demonstration here will be for the equal weighted average; however, it would be straightforward to extend the presentation to include the weights in Equation (2.11). Note that the mean in Equation 2.10 or 2.11 are linear combinations of the available data.

Calculation of the second-order moment of this linear combination amounts to calculating the covariance between the ith and j^{th} RV, that is Z_i and Z_j, i, $j=1,\ldots,n$. This covariance is denoted as $\{C_{ij}, i=1,\ldots,n; j=1,\ldots,n\}$ and is formally defined as:

$$\begin{aligned} Cov\{Z_i, Z_j\} &= C_{ij} \\ &= E\{(Z_i - m_i)(Z_j - m_j)\} \\ &= E\{Z_i Z_j\} - m_i m_j \end{aligned} \quad (2.12)$$

where m_i is the mean of the RV Z_i. Note the similarity between the definition of the covariance in Equation (2.12) and the variance in Equation (2.3). We can think of the variance as the covariance between a RV with itself; in fact, C_{ii} is the variance of the ith variable, Z_i. In cases where Z_i and Z_j have the same mean, then $m_i = m_j = m$ and Equation (2.12) reduces to

$$C_{ij} = E\{Z_i Z_j\} - m^2 \quad (2.13)$$

Problem

(i) Express the variance of the equal weighted mean of the data, $\bar{z}$, in terms of the covariance.

(ii) Rewrite the variance in (i) under the assumption of independence between the data.

Solution Plan

(i) Solve for $Var\{\bar{z}\}$ in terms of C_{ij} noting that $C_{ij} = E\{z_i z_j\} - \bar{z}^2$

(ii) Repeat the solution, noting that in the presence of independence, $C_{ij} = 0, \ \forall i \neq j$ and $C_{ii} = \sigma^2$.

Solution

(i) Write the expression for the variance based on Equation (2.3) with the RV $\bar{Z}$:

$$
\begin{aligned}
Var\{\bar{z}\} &= E\{\bar{z}^2\} - \bar{z}^2 \\
&= E\left\{\left(\frac{1}{n}\sum_{i=1}^{n} z_i\right)^2\right\} - \bar{z}^2 \\
&= \frac{1}{n^2}\sum_{i=1}^{n}\sum_{j=1}^{n} E\{z_i z_j\} - \bar{z}^2 \\
&= \frac{1}{n^2}\sum_{i=1}^{n}\sum_{j=1}^{n} C_{ij}
\end{aligned}
\tag{2.14}
$$

The expected value operator, $E\{\cdot\}$, is a probability weighted average. It is an integral over all possible values and, like the integral operator, is linear. The expected value of a constant times a random variable is the constant times the expected value of the random variable: $E\{aZ\} = aE\{Z\}$. The expected value of a sum is the sum of expected values. Note also that $E\{(aZ)^2\} = a^2E\{Z^2\}$.

(ii) The assumption of data independence simplifies the solution to (i) whereby only certain components of the covariance, C_{ij}, must be considered. Specifically, $C_{ij} = 0$ where $i \neq j$. Note further that C_{ii} is equivalent to the variance of the ith RV.

$$
\begin{aligned}
Var\{\bar{z}\} &= E\{\bar{z}^2\} - \bar{z}^2 \\
&= \frac{1}{n^2}\left(\sum_{i=1}^{n}\sum_{\substack{j=1 \\ j\neq i}}^{n} C_{ij} + \sum_{i=1}^{n} C_{ii}\right) \\
&= \frac{nC_{ii}}{n^2} \\
&= \frac{\sigma^2}{n}
\end{aligned}
\tag{2.15}
$$

Remarks

This simple exercise demonstrates several key concepts. First, correlated data results in higher uncertainty in the mean due to redundancy between the data. In these cases, the variance of the mean is the average of the covariance between the data pairs. This finding is the basis for linear estimation in the presence of correlation between the data. Second, in the case of independence between the data, the result is predicted by the Central Limit Theorem and is a classical result found in statistics. The importance of the Central Limit Theorem will be revisited in Chapter 4.

The important principle of stationarity is also revealed in this exercise. Stationarity is the choice of pooling data for common analysis: The initial choice of the n data. Stationarity is also the choice of location-dependence of statistical parameters. Equal variance for all data is part of our decision of stationarity. Pooling data to calculate covariances is another part of our decision of stationarity.

Similar demonstrations of this linear combination problem can be found in Armstrong (1998, pp. 32–35) and Chilès and Delfiner (1999, p. 158)

2.3 STANDARDIZATION AND PROBABILITY INTERVALS

Learning Objectives

Many of our inference tools rely on the Gaussian (or normal) distribution. For this reason, this exercise aims to improve understanding and comfort level with the statistics and units of the normal distribution.

Background and Assumptions

Almost all our data are nonstandard, that is, the mean is nonzero and the variance is not equal to 1.0. In many cases, standardizing the variable simplifies the mathematics of a problem by removing excess operations. The term "standardization" then refers to changing the units of the data by centering it to a mean of 0.0 and scaling the variance to be 1.0. This result is easily achieved by

$$y = \frac{z - \mu}{\sigma} \tag{2.16}$$

where Y is the standardized variable, Z is the nonstandard (or original) variable, μ is the mean of Z, and σ is the standard deviation of Z. This simple transformation can be reversed by reorganizing the equation as:

$$z = \mu + \sigma y \tag{2.17}$$

These standardization equations can be applied to any distribution; however, the shape of the distribution is not changed.

The Gaussian distribution is one of the most commonly used distributions because of its simplicity. In its univariate form, it only requires two parameters, the mean and the variance, to fully define the distribution:

$$f(z) = \frac{1}{\sigma\sqrt{2\pi}} e^{-\frac{1}{2}\left(\frac{z-\mu}{\sigma}\right)^2}$$

where μ and σ are the mean and standard deviation respectively. Further, if Z is a standard Gaussian variable, then the probability density function (pdf) is simplified

$$f(z) = \frac{1}{\sqrt{2\pi}} e^{-\frac{z^2}{2}}$$

While the pdf is analytically defined, the cumulative frequency distribution (cdf) has no closed-form analytical solution. For this reason, probability tables exist at the back of most statistics and probability textbooks tabulating the cdf values for different Gaussian distributions.

The previous exercises focused on two important statistics: mean and variance. Other statistics are of interest when we consider how uncertainty is reported. One approach to report uncertainty is to consider the pth probability interval, which is a range of values whereby the minimum and maximum of this range corresponds to the $(1-p)/2$ and $(1+p)/2$ quantile, respectively. For instance, the 90% probability interval is reported using the 5th quantile and the 95th quantile. In general, large probability intervals indicate a large distribution of uncertainty.

Problem

Consider 10 data that are available over a two-dimensional (2D) area of interest that is nominally 50 x 50 distance units square (see Table 2.1). Further, assume that the data is known to be a normal distribution with a mean of 12.0 and a variance of 16.0.

(i) Calculate the 95% centered probability interval for the mean, that is, the 0.025 quantile and the 0.975 quantile of the theoretical distribution for the mean.

(ii) Transform the data to be standard normal and recalculate the 95% probability interval.

(iii) Consider now that the data are spatially dependent, and this dependence is characterized by a correlation function specified as $C(\mathbf{h}) = \exp(-0.2\mathbf{h})$ where

h is calculated as the distance between the samples. Note that this function applies to the standard variable, that is, C(**0**)=1.0 (the variance of a standard variable). These values would have to be multiplied by 16 to establish the actual covariance values. Calculate the variance of the mean in this case and calculate the 95% centered probability interval.

Table 2.1 Available Samples from a Gaussian Reference Distribution

Data	Easting	Northing	Variable
1	20.5	1.5	12.9
2	6.5	25.5	11.1
3	9.5	19.5	9.2
4	40.5	7.5	16.4
5	28.5	11.5	8.4
6	14.5	21.5	14.8
7	11.5	46.5	19.3
8	38.5	28.5	14.0
9	7.5	12.5	13.5
10	13.5	24.5	13.5

Solution Plan

(i) The probability interval of a Gaussian variable can be obtained by integration of the Gaussian pdf; however, this job is tedious given that tables and many programs exist to calculate this for us. Most spreadsheet applications have a function that can output the Gaussian (or normal) inverse given a probability value, mean, and variance. Note that the Gaussian inverse value is equivalent to the quantile, that is, $G(y)=p$, thus $y=G^{-1}(p)$ where $G(\cdot)$ represents the Gaussian distribution, p represents the probability, and y is the corresponding quantile.

(ii) Use Equation (2.16) to transform each of the 10 values to standard normal. Note that the equation requires the standard deviation, not the variance. Calculate the probability interval in a similar fashion to Part (i) above, making sure the mean and variance are changed accordingly.

(iii) Use the results of Problem 2.2 Part (i) to calculate the variance of the mean (i.e., the linear combination) and calculate each of the covariance terms in that linear summation. This first requires a calculation of the distance between all samples that will yield a 10 x 10 matrix of distances. The reader can perform this calculation by hand (or some spreadsheet application), or alternatively write a small program to do this. Apply the given covariance function and sum over all data pairs. Once the variance of the mean is determined, use this variance to recalculate the probability interval.

Solution

(i) Using MS Excel, we can calculate the 0.025 quantile as 4.16 and the 0.975 quantile as 19.84. Thus, the 95% probability interval is [4.16, 19.84] centered about the mean of 12.

(ii) Applying Equation (1.16) yields the standardized values shown in Table 2.2. Once again, using MS Excel, we can calculate the 95% probability interval as [−1.96, +1.96]. Note further that we could have also applied the standardization equation to the results from part (i) to obtain the same interval.

Table 2.2 Standardized Values for the Original Gaussian Dataset.

Data	Easting	Northing	Variable	Standardized Variable
1	20.5	1.5	12.9	0.225
2	6.5	25.5	11.1	−0.225
3	9.5	19.5	9.2	−0.7
4	40.5	7.5	16.4	1.1
5	28.5	11.5	8.4	−0.9
6	14.5	21.5	14.8	0.7
7	11.5	46.5	19.3	1.825
8	38.5	28.5	14	0.5
9	7.5	12.5	13.5	0.375
10	13.5	24.5	13.5	0.375

(iii) First calculate the 10 x 10 matrix of distances between each pair of data, then apply the covariance function, $C(\mathbf{h}) = \exp(-0.2\mathbf{h})$ to each of the distances in the matrix (see Table 2.3).

Average all the covariance values in the 10 x 10 covariance matrix to obtain the variance of the mean of 0.155. The corresponding probability interval is [−0.77, 0.77]. Note that in the case of independence, this same variance is 0.10 and would yield a smaller 95% probability interval. These calculations apply to the standardized case; to obtain similar statistics for the nonstandard case (i.e., the original case with a variance of 16), multiply the covariances in the matrix by 16. Note further that in the case of standardized variables, the covariance is the same as the correlation.

Table 2.3 Distance Matrix (a) between 10 Data and the Corresponding Covariance Matrix (b).

(a)

Distance	1	2	3	4	5	6	7	8	9	10
1	0	27.785	21.095	20.881	12.806	20.881	45.891	32.45	17.029	24.042
2	27.785	0	6.708	38.471	26.077	8.944	21.587	32.14	13.038	7.071
3	21.095	6.708	0	33.242	20.616	5.385	27.074	30.364	7.28	6.403
4	20.881	38.471	33.242	0	12.649	29.53	48.6	21.095	33.377	31.906
5	12.806	26.077	20.616	12.649	0	17.205	38.91	19.723	21.024	19.849
6	20.881	8.944	5.385	29.53	17.205	0	25.179	25	11.402	3.162
7	45.891	21.587	27.074	48.6	38.91	25.179	0	32.45	34.234	22.091
8	32.45	32.14	30.364	21.095	19.723	25	32.45	0	34.886	25.318
9	17.029	13.038	7.28	33.377	21.024	11.402	34.234	34.886	0	13.416
10	24.042	7.071	6.403	31.906	19.849	3.162	22.091	25.318	13.416	0

(b)

Covariance	1	2	3	4	5	6	7	8	9	10
1	1.0000	0.0039	0.0147	0.0154	0.0772	0.0154	0.0001	0.0015	0.0332	0.0082
2	0.0039	1.0000	0.2614	0.0005	0.0054	0.1672	0.0133	0.0016	0.0737	0.2431
3	0.0147	0.2614	1.0000	0.0013	0.0162	0.3406	0.0045	0.0023	0.2332	0.2779
4	0.0154	0.0005	0.0013	1.0000	0.0797	0.0027	0.0001	0.0147	0.0013	0.0017
5	0.0772	0.0054	0.0162	0.0797	1.0000	0.0320	0.0004	0.0194	0.0149	0.0189
6	0.0154	0.1672	0.3406	0.0027	0.0320	1.0000	0.0065	0.0067	0.1022	0.5313
7	0.0001	0.0133	0.0045	0.0001	0.0004	0.0065	1.0000	0.0015	0.0011	0.0121
8	0.0015	0.0016	0.0023	0.0147	0.0194	0.0067	0.0015	1.0000	0.0009	0.0063
9	0.0332	0.0737	0.2332	0.0013	0.0149	0.1022	0.0011	0.0009	1.0000	0.0683
10	0.0082	0.2431	0.2779	0.0017	0.0189	0.5313	0.0121	0.0063	0.0683	1.0000

Remarks

This simple exercise demonstrates the ease of moving between standardized to nonstandardized units. Working with the Gaussian distribution requires familiarity with the units of the distribution, and that these units are symmetric about the mean. For a standardized Gaussian distribution, one should expect the values to range from approximately -3 to $+3$; note that 95% of the values range from -1.96 to $+1.96$ (as given by the 95% probability interval). Values beyond the ± 3 range may indicate outliers and require further investigation.

This exercise has also introduced the concept of spatial correlation through the use of $C(\mathbf{h})$. Although the calculation and fitting of $C(\mathbf{h})$ has not been explained, it is important for students of geostatistics to become comfortable considering all data in a spatial context with spatial correlation.

CHAPTER 3

Obtaining Representative Distributions

Data in earth science applications are rarely collected uniformly over the site of interest. The focus is usually aimed at ascertaining the size, orientation, and concentration of a natural resource, such as a mineral deposit or a petroleum accumulation, or the extent of some environmental contaminant. Data are often expensive to collect and it is neither optimal nor feasible to collect data uniformly over the entire site being characterized. Moreover, the available data may be production locations that are not chosen with the goal of informing spatial statistics. Finally, in certain cases, portions of the site are less accessible leading to preferential sampling. For these reasons, the data available for geostatistical modeling cannot be considered equally in the construction of probability distributions and other statistics.

Declustering assigns weights to the available data based on the configuration of data. Data in sparsely sampled areas are given more weight and clustered data are given less weight. The data values are not changed; they are simply given more or less influence based on their spacing. Olea (2007) gives a good overview of different declustering methods.

Debiasing uses a secondary variable, such as geophysical measurements or a geological trend, to correct the global statistics (Gringarten et al., 2000; Deutsch, 2002; Pyrcz and Deutsch, 2003). Once again, the data values are not changed; a relationship with a secondary data is exploited to establish representative statistics.

Some techniques do not require a representative distribution. Variogram calculation, inverse distance estimation, and ordinary kriging do not require a representative global distribution to be specified; however, most geostatistical

Solved Problems in Geostatistics. By O. Leuangthong, K.D. Khan, and C.V. Deutsch

techniques require a representative distribution. Uncertainty assessment and simulation are particularly sensitive to a reasonable assessment of the representative distribution.

The problems in this chapter focus on the challenge of determining a representative global distribution for subsequent calculations. The first problem considers the problem of assigning declustering weights to data based on the geometric arrangement of the data and sample spacing. The second one looks at the case where samples from a portion of the distribution are not available and we must debias this distribution. The third problem compares four of the most widely used declustering methods: polygonal, cell, kriging, and inverse distance.

3.1 BASIC DECLUSTERING

Learning Objectives

This problem explores the concept of assigning weights to data for a representative histogram and summary statistics. Differentially weighted data allow the computation of "corrected" statistics, as opposed to 'naïve' statistics calculated from equal-weighted data.

Background and Assumptions

Weights are assigned to the data to make the global statistics representative. Assigning equal weights to the available data is often naïve and leads to a systematic bias in the global distribution and summary statistics. Figure 3.1 shows an equal weighted and a weighted distribution of the same five data. The weights to the two highest data are reduced in the left illustration; the assumption is that those two highest values were sampled in close proximity to one another; therefore, they receive less weight than the other three.

Declustering modifies the entire distribution (Figure 3.1) and summary statistics, such as the mean and standard deviation. The mean of a random variable (RV) is a probability weighted average of the data values. Empirically, this is calculated as:

$$\overline{z} = \frac{\sum_{i=1}^{n} w_i z_i}{\sum_{i=1}^{n} w_i} \tag{3.1}$$

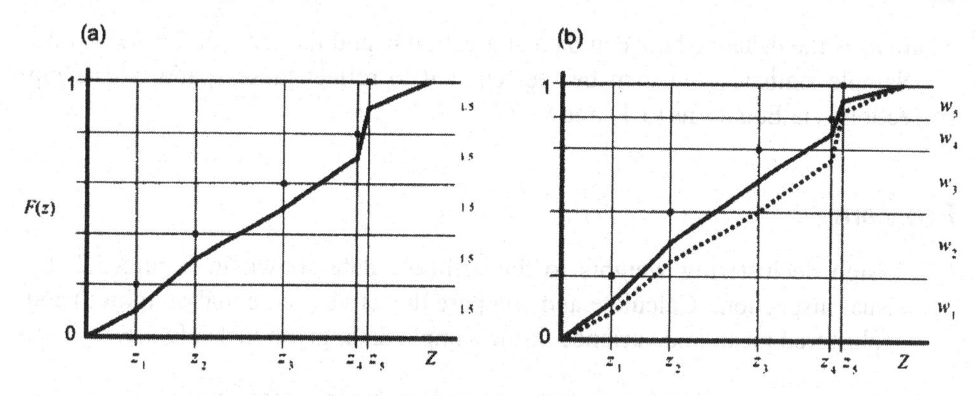

Figure 3.1 Nonparametric cumulative distribution function calculated from five data with (a) equal weights and (b) unequal weights.

where w_i is the weight assigned to i^{th} data value $z_i, i = 1,...,n$. In the naïve case, where we do not consider declustering, each data is assigned the same (or equal) weight. This weight is *1/n*, where *n* is the number of data. In a similar manner to calculating the mean, the sample standard deviation can be calculated as:

$$\sigma = \sqrt{\frac{\sum_{i=1}^{n} w_i \left(z_i - \bar{z}\right)^2}{\sum_{i=1}^{n} w_i}} \tag{3.2}$$

where $\bar{z}$ comes from Equation (3.1).

The declustering weights ($w_i, i=1,...n$) must be nonnegative and sum to 1.0. Virtually all software applications will restandardize the weights automatically; therefore, it is often convenient to leave the weights in the units of area of influence or relative to *n*, that is, the sum of the weights is the number of data *n*.

Declustering methods are primarily based on the geometrical configuration of the data; they do not generally consider the data values. There are various methods to assign weights to the data to obtain globally representative statistics. One such approach is to 'expertly' assign these weights via a visual inspection; however, this works in only the simplest of cases and is rarely applicable to large scale, complex geological sites.

Another approach is to consider a nearest-data weighting scheme, whereby samples are weighted based on how close the nearest *n* samples are

$$w_j = \sum_{\substack{i=1 \\ i \neq j}}^{n} h_{ij} \tag{3.3}$$

where h_{ij} is the distance between data at location $\mathbf{u}_i$ and the location of interest $\mathbf{u}_j$.

Sample statistics can then be recalculated to reflect these spatial clustering corrections via Equations (3.1) and (3.2).

Problem

(i) Assign declustering weights to the drillhole data shown in Figure 3.2. by visual inspection. Calculate and compare the naive (i.e., equal-weighted) and declustered mean and variance of the sample data, given in Table 3.1.

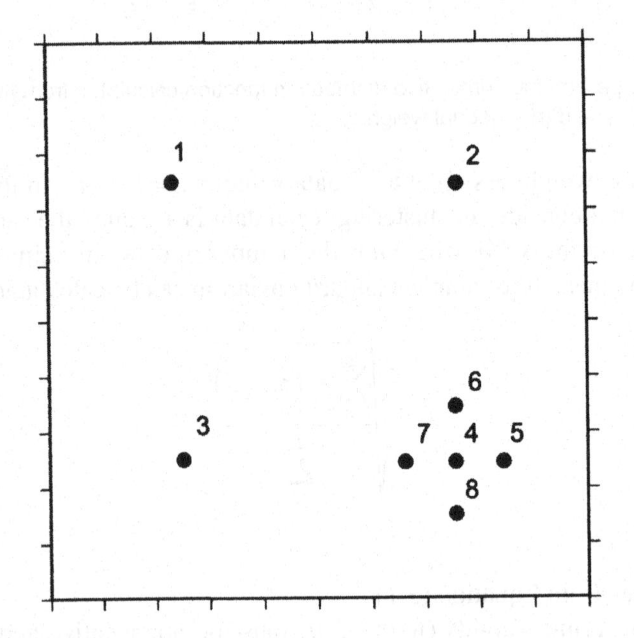

Figure 3.2 Drillhole locations for Part (i) Note that distance units are not important to assigning declustering weights for this simple configuration.

Table 3.1 Exploration Sample Data for Part (i)

DH No.	Grade
1	2.0
2	1.8
3	2.2
4	8.2
5	7.8
6	7.9
7	8.4
8	8.8

(ii) Consider a gold-rich deposit with a lease limit that measures 1200 m x 1200 m. For preliminary exploration purposes, the lease has been sampled at 15 locations shown in Figure 3.3 and tabulated in Table 3.2. Perform declustering on the sample data using the distance to the nearest data criterion in Equation (3.3), considering only the three closest data for this weighting (i.e., $n=3$). Calculate the corresponding declustered mean grade and standard deviation. Also, calculate and compare the equal weighted mean and variance to this declustered mean and variance.

Solution Plan

(i) A visual inspection reveals some regularity in the spacing of the data, particularly if we consider quadrants of this field. Assign declustering weights based on the spacing of sampled data, and then calculate the summary statistics. You may wish to consider a weighting scheme in which the sum of the weights equals 1.0 (although this is not required, some may find it useful to work with this constraint).

(ii) Calculate the spacing between pairs of drillholes using a spreadsheet, and determine the n closest data to each hole as a local search criterion. In this example, $n = 3$. Calculate the weight to assign each drillhole location based on Equation (3.3). Calculate the declustered mean and variance of the grade over the lease.

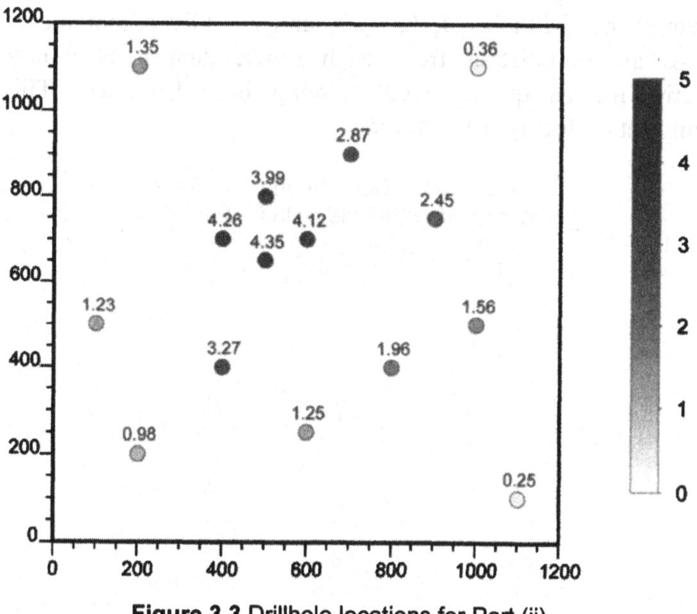

Figure 3.3 Drillhole locations for Part (ii).

Table 3.2 Exploration Sample Data for Part (ii).

Drillhole Number	East (m)	North (m)	Grade
1	100	500	1.23
2	500	650	4.35
3	1000	1100	0.36
4	1000	500	1.56
5	700	900	2.87
6	600	250	1.25
7	400	400	3.27
8	400	700	4.26
9	200	200	0.98
10	500	800	3.99
11	200	1100	1.35
12	800	400	1.96
13	1100	100	0.25
14	600	700	4.12
15	900	750	2.45

Solution

(i) If we break the area up into quadrants, then it is reasonable to assign a quarter of the weight to each of drillholes 1, 2, and 3. The last quarter must be "shared" by drillholes 4, 5, 6, 7, and 8. Given that these latter five drillholes are equidistant from each other, then it is fair to consider distributing the last quarter evenly among these drillholes. This gives the following set of declustering weights:

Table 3.3 Declustering Weights Assigned to Sample Data for Part (i)

DH No.	Grade	Weight
1	2.0	0.25
2	1.8	0.25
3	2.2	0.25
4	8.2	0.05
5	7.8	0.05
6	7.9	0.05
7	8.4	0.05
8	8.8	0.05

The declustered sample mean and standard deviation are calculated using Equations (3.1) and (3.2), respectively, to yield the following results:

Statistic	Equal-weighted	Declustered
$\bar{z}$	5.89	3.56
σ	3.03	2.70

(ii) The distance, h_{ij}, between two data is calculated as:

$$h_{ij} = \sqrt{\left(x_j - x_i\right)^2 + \left(y_j - y_i\right)^2}$$

Table 3.4 shows the distance between data pairs for the first three drillholes using the drillhole coordinates in Table 3.2. A complete table should be constructed for the remaining drillholes. Given the distance between data and using Equation (3.3), the weights assigned to drillholes 1, 2, and 3 are given:

$$w_1 = 993.01$$
$$w_2 = 373.61$$
$$w_3 = 1290.25$$

Table 3.4 Distance between Drillholes 1–3 and All Other Drillholes Over the Entire Region.*

Drillhole (n)	Distance to DH1	Distance to DH2	Distance to DH3
1	0	427.2	1081.7
2	427.2	0	672.7
3	1081.7	672.7	0
4	900	522	600
5	721.1	320.2	360.6
6	559	412.3	939.4
7	316.2	269.3	922
8	360.6	111.8	721.1
9	316.2	540.8	1204.2
10	500	150	583.1
11	608.3	540.8	800
12	707.1	390.5	728
13	1077	813.9	1005
14	538.5	111.8	565.7
15	838.2	412.3	364

*The closest three data points to the respective drillhole are indicated by shaded cells.

The declustered sample mean and standard deviation using the complete set of weights are

Statistic	Equal-weighted	Declustered
$\bar{z}$	2.28	1.80
s	1.39	1.74

Remarks

Although the weighting scheme implemented here is not exactly a nearest neighbor declustering (see Problem 3.3), it does assign greater weight to data in sparsely sampled areas while smaller weights are given to data in densely sampled regions. This result is consistent with the idea of assigning weights based on information content; further away data inform larger regions, hence greater weight is given to these samples.

Note that declustering assumes that the entire range of the population has been sampled; that is, despite preferential sampling practices, low, median, and high valued regions were all sampled

3.2 DEBIASING WITH BIVARIATE GAUSSIAN DISTRIBUTION

Learning Objectives

Debiasing is required when one or more portion(s) of the true distribution are unsampled. For example, when only high-valued regions are sampled, the lower portion of the underlying true distribution is not represented at all. This difference is important from declustering where low data are available, but unfairly represented. The goal of this problem is to correct the bias in a sample distribution of a primary attribute due to unavailability of a portion of the underlying population.

Background and Assumptions

We are often interested in debiasing, that is, deriving a representative distribution for an attribute from some other source of data (i.e., a secondary attribute and associated calibration data). Consider two RVs: Y represents the attribute of interest and X represents an exhaustive secondary variable, that is, one that is available over the entire study area. The joint probability space between Y and X is assumed to follow a bivariate Gaussian distribution, where ρ is the correlation coefficient between the X and Y random variables:

$$f_{xy}(x,y)=\frac{1}{2\pi\sigma_X\sigma_Y\sqrt{1-\rho^2}}e^{\left[\frac{1}{(1-\rho^2)}\left(\frac{(x-m_x)^2}{2\sigma^2_X}+\frac{(y-m_y)^2}{2\sigma^2_Y}-\rho\frac{(x-m_x)(y-m_y)}{\sigma_X\sigma_Y}\right)\right]} \tag{3.4}$$

If X and Y are standard Gaussian, then Equation (3.4) simplifies to

$$f_{xy}(x,y)=\frac{1}{2\pi\sqrt{1-\rho^2}}e^{\left[\frac{1}{2(1-\rho^2)}(x^2-2\rho xy+y^2)\right]} \tag{3.5}$$

By using the exhaustive secondary data represented by X to debias the original distribution of Y given a joint distribution, $f_{xy}(x,y)$, requires that we invoke Bayes theorem. Recall that Bayes theorem relates a conditional event, $A|B$ (A given B), to the joint (A and B) and marginal (B) events:

$$P(A\,|\,B)=\frac{P(A \text{ and } B)}{P(B)} \tag{3.6}$$

The event can be taken as the distributions and Equation (3.6) becomes

$$f_{y|x}(y)=\frac{f_{xy}(x,y)}{f_x(x)} \tag{3.7}$$

where $f_{y|x}(y)$ is the conditional distribution of Y given that X=x. Thus, the joint distribution can be determined by rearranging Equation (3.7):

$$f_{xy}(x,y)=f_{y|x}(y)\cdot f_x(x) \tag{3.8}$$

To determine the marginal distributions, $f_x(x)$ and $f_y(y)$, we can integrate over the joint pdf, $f_{x,y}(x,y)$. For example, $f_y(y)$ is obtained by integrating $f_{x,y}(x,y)$ over X:

$$f_y(y)=\int_X f_{x,y}(x,y)dx \tag{3.9}$$

Substituting Equation (3.8) into Equation (3.9) yields

$$f_y(y) = \int_X f_{y|x}(y) \cdot f_x(x)dx \tag{3.10}$$

By using these relationships, we can analytically or numerically perform debiasing through a Gaussian model. In this problem, an analytical method is devised.

Problem

(i) Develop a methodology for debiasing the pdf of a RV Y, $f_{\hat{Y}}(y)$, given the following information (see Figure 3.4): (a) the joint distribution required for calibration information is bivariate Gaussian with parameters $m_Y, \sigma_Y, m_X, \sigma_X$, and $\rho_{X,Y}$ derived from available information (i.e., likely biased); and (b) a representative Gaussian distribution for the X variable is available, which is specified by $m_{\hat{X}}, \sigma_{\hat{X}}$.

(ii) Given that the bivariate distribution is defined by $m_Y=10$, $\sigma_Y=8$, $m_X=250$ and $\sigma_X=50$, implement debiasing for various correlation coefficients, $\rho_{X,Y}$ = −0.90, −0.50, 0.0, 0.50, 0.90. The representative distribution of the secondary variable X is Gaussian with $m_{\hat{X}} = 200$, $\sigma_{\hat{X}} = 40$. For each correlation coefficient, calculate the corresponding mean, $m_{\hat{Y}}$, and standard deviation, $\sigma_{\hat{Y}}$, of the debiased distribution.

Solution Plan

(i) Debiasing a distribution requires calibration data; for this problem, this data comes in the form of an exhaustive secondary RV, X. However, we must also know the relationship between X and Y so that we can take advantage of the exhaustive information. Fortunately, this is also known and it follows a bivariate Gaussian distribution that is defined in Equation (3.4). The idea then is to develop a form for the debiased distribution, $f_{\hat{Y}}(y)$, that draws from the joint distribution, $f_{XY}(x,y)$ and the representative $f_{\hat{X}}(x)$. *Hint*: use Bayes' law as shown in Equations (3.7)–(3.10).

(ii) Using the debiased distribution from Part (i), the mean, $m_{\hat{Y}}$, and standard deviation, $\sigma_{\hat{Y}}$, can be derived by using Equations (2.2) and (2.3). Once the equations to calculate the mean and standard deviation are obtained, it is possible to substitute in the means, standard deviations, and correlation coefficient to obtain a value.

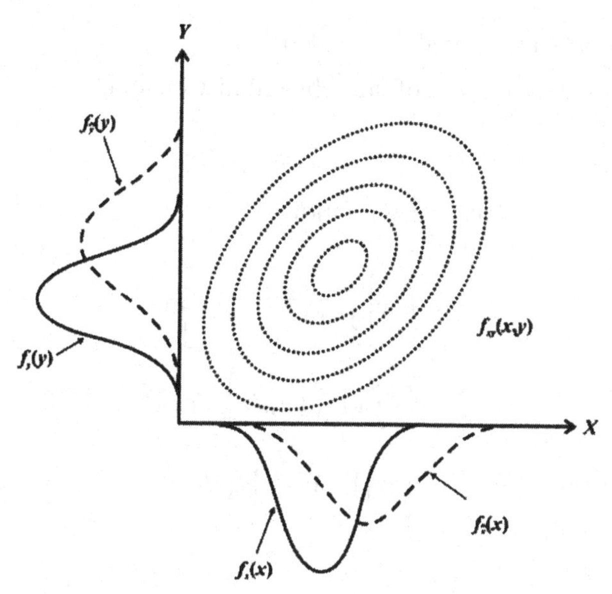

Figure 3.4 Schematic Illustration of debiasing with bivariate Gaussian distribution: dashed lines correspond to the debiased distributions, solid lines correspond to the original data distribution.

Solution

(i) Deriving the debiased distribution, $f_{\hat{Y}}(y)$, is possible because we have a secondary variable, $\hat{X}$, with representative statistics, $m_{\hat{X}}$ and $\sigma_{\hat{X}}$, and the calibration relationship, namely, a bivariate Gaussian pdf given by Equation (3.4). We can then use Bayes' theorem to determine the conditional distribution of $Y|X$; note that this distribution only requires the calibration relationship.

$$f_{Y|X}(y) = \frac{f_{X,Y}(x,y)}{f_X(x)} \tag{3.11}$$

Given the conditional distribution of Y given X ($Y|X$), we can now determine the debiased marginal distribution of Y, $f_{\hat{Y}}(y)$, by weighting each of the conditional distributions ($Y|X$) by the representative distribution of X, $f_{\hat{X}}(x)$.

$$f_{\hat{Y}}(y) = \int_X f_{Y|X}(y) f_{\hat{X}}(x)dx \tag{3.12}$$

and $f_{\hat{X}}(x)$ is Gaussian with $m_{\hat{X}}$ and $\sigma_{\hat{X}}$.

(ii) Solving for the mean, $m_{\hat{Y}}$, of the debiased distribution:

$$\begin{aligned} m_{\hat{Y}} &= \int_Y y\, f_{\hat{Y}}(y) dy \\ &= \int_Y y \left[\int_X f_{Y|X}(y) f_{\hat{X}} dx \right] dy \\ &= \int_X \left[\int_Y y f_{Y|X}(y) dy \right] f_{\hat{X}}(x) dx \\ &= \int_X E\{Y \mid X = x\} f_{\hat{X}} dx \end{aligned}$$

The conditional expectation $E\{Y \mid X = x\}$ of a bivariate Gaussian distribution can be determined by performing the operations above or we can look it up in most statistics textbooks. It is given as $E\{Y \mid X = x\} = m_Y + \rho \frac{\sigma_Y}{\sigma_X}(x - m_X)$, thus the debiased mean, $m_{\hat{Y}}$, becomes

$$\begin{aligned} m_{\hat{Y}} &= \int_X \left[m_Y + \rho \frac{\sigma_Y}{\sigma_X}(x - m_X) \right] f_{\hat{X}}(x) dx \\ &= m_Y + \rho \frac{\sigma_Y}{\sigma_X} \left[\int_X f_{\hat{X}}(x) dx - m_X \right] \\ &= m_Y + \rho \frac{\sigma_Y}{\sigma_X} \left(m_{\hat{X}} - m_X \right) \end{aligned} \tag{3.13}$$

Following the same logic above and using Equation (2.3), we can solve for the variance of the debiased distribution:

$$\begin{aligned}
\sigma_{\hat{Y}}^2 &= E\left\{\hat{Y}^2\right\} - \left[E\left\{\hat{Y}\right\}\right]^2 \\
&= E\left\{\hat{Y}^2\right\} - m_{\hat{Y}}^{\ 2} \\
&= \int_Y y^2 \left[\int_X f_{Y|X}\left(y \mid X = x\right) f_{\hat{X}}(x)dx\right] dy - m_{\hat{Y}}^{\ 2} \\
&= \int_X f_{\hat{X}}(x) \left[\int_Y y^2 f_{Y|X}\left(y \mid X = x\right) dy\right] dx - m_{\hat{Y}}^{\ 2} \\
&= \int_X f_{\hat{X}}(x) \left[E\{Y^2 \mid X = x\}\right] dx - m_{\hat{Y}}^{\ 2} \\
&= \int_X \left[Var\left\{Y \mid X = x\right\} + \left(E\left\{Y \mid X = x\right\}^2\right)\right] f_{\hat{X}}(x)dx - m_{\hat{Y}}^{\ 2}
\end{aligned}$$

Note that we use $Var\{Y|X{=}x\}{=}E\{Y^2|X{=}x\}{-}E\{Y|X{=}x\}^2$ in the simplification above. Note also that the above equation requires the conditional variance of $Y|X{=}x$, and once again, we can look this up in a statistics textbook to find that $Var\left\{Y \mid X = x\right\} = \sigma_Y^2\left(1 - \rho^2\right)$, thus:

$$\begin{aligned}
\sigma_{\hat{Y}}^2 &= \int_X \left[\sigma_Y^2\left(1 - \rho^2\right) + \left(m_Y + \rho\frac{\sigma_Y}{\sigma_X}(x - m_X)\right)^2\right] f_{\hat{X}}(x)dx - m_{\hat{Y}}^{\ 2} \\
&= \sigma_Y^2\left(1 - \rho^2\right) + \left(m_Y^{\ 2} - m_{\hat{Y}}^{\ 2}\right) + 2m_Y\rho\frac{\sigma_Y}{\sigma_X}\left(m_{\hat{X}} - m_X\right) \\
&\quad + \rho^2\frac{\sigma^2_{\ Y}}{\sigma^2_{\ X}}\left(E\left\{\hat{X}^2\right\} - 2m_X m_{\hat{X}} + m_X^{\ 2}\right)
\end{aligned}$$

Substitution via Equation (3.13), expansion and cancellation of terms yields

$$\begin{aligned}
\sigma^2_{\ \hat{Y}} &= \sigma^2_{\ Y}\left[1 - \rho^2 + \rho^2\frac{\sigma^2_{\ \hat{X}}}{\sigma^2_{\ X}}\right] \\
&= \sigma^2_{\ Y}\left[1 + \rho^2\left(\frac{\sigma^2_{\ \hat{X}}}{\sigma^2_{\ X}} - 1\right)\right] \qquad (3.14) \\
\sigma_{\hat{Y}} &= \sigma_Y\sqrt{1 + \rho^2\left(\frac{\sigma^2_{\ \hat{X}}}{\sigma^2_{\ X}} - 1\right)}
\end{aligned}$$

Now that we have an expression for both the debiased mean and variance, we can substitute in the given values $m_Y = 10$, $\sigma_Y = 8$, $m_X = 250$, $\sigma_X = 50$, $m_{\hat{X}} = 200$, and $\sigma_{\hat{X}} = 40$. For different correlation coefficients, Table 3.5 gives the corresponding mean and standard deviation.

Table 3.5 Debiased Mean and Standard Deviation for Different Correlation Coefficients.

$\rho_{X,Y}$	$m_{\hat{Y}}$	$\sigma_{\hat{Y}}$
0.9	2.8	6.73
0.5	6.0	7.63
0	10.0	8.00
−0.5	14.0	7.63
−0.9	17.2	6.73

Remarks

In the case of independence ($\rho_{X,Y}$=0.0), $m_{\hat{Y}}$ and $\sigma_{\hat{Y}}$ are equal to m_Y and σ_Y. The debiasing information contained in the parameters of $f_{\hat{X}}(x)$ has no effect on the marginal $f_Y(y)$.

The symmetry of the bivariate Gaussian pdf (**Error! Reference source not found.**) is apparent in the calculation results. The mean, $m_{\hat{Y}}$, shifts higher or lower from the biased mean, m_Y, depending on the sign on the correlation coefficient in a symmetrical fashion, as given by Equation (3.13). With any correlation, there is a proportional decrease in the variance of the debiased marginal distribution, $f_{\hat{Y}}(y)$, from the biased marginal distribution, $f_Y(y)$, based on Equation (3.14).

This problem is more theoretical than practical. In practice, the bivariate distribution between the variable of interest and the secondary data will be fitted non parametrically with all available data and additional control points if necessary. The additional control points or pseudodata would constrain the extrapolation into regions where the controlling primary variable has not been sampled. Deutsch (2002, pp. 59–63) discusses this in further detail.

3.3 COMPARISON OF DECLUSTERING METHODS

Learning Objectives

This problem compares different algorithms for computing declustering weights used for debiasing. Four declustering approaches are considered in order to compare the weighting scheme and to understand the reasons for the possibly different weights assigned to the same set of samples.

Background and Assumptions

There are four well-documented means of assigning declustering weights (Isaaks and Srivastava, 1989, pp. 237–248; Goovaerts, 1997, pp. 77–82; Deutsch, 2002, pp 50–62): polygonal, cell, kriging, and inverse distance.

Polygonal declustering is applied where the data or data collection sites are irregularly spaced, such as meteorological stations. In this case, there is no underlying natural or pseudo regular grid to apply a cell-based declustering technique. Polygonal declustering is generally described and applied in 2D, although it is possible, though much more complicated, to construct the polygons in 3D. A basic method in 2D is to connect all nearest-neighbor pairs of data locations, and then extend the perpendicular bisectors of these connections to form closed polygons, one surrounding each datum location. As the data spacing decreases, polygon size around the data decreases. The relative area, A, of the polygon (p) around the ith datum is then assigned as a declustering weight:

$$w_i^{(p)} = \frac{A_i}{\sum_{j=1}^{N} A_j} \tag{3.15}$$

The configuration of the domain boundary may strongly influence the weights assigned to the data on the spatial periphery of the sample set.

Cell declustering considers equal-sized rectilinear cells with sides aligned with the coordinate system so they are naturally definable in 3D Cartesian space. The weights are assigned based on the number of data occupying the same cell, so the ith, $i=1,\dots,n$, clustered data receives the following weight:

$$w_i^{(c)} = \frac{1}{n_j \cdot J} \tag{3.16}$$

where J is the number of cells occupied by one or more data, and n_j is the number of data in the jth cell, $j=1, ..., J, J \leq n$. Cell declustering works best by trying many different cell sizes. Application of cell declustering may require some number of origin offsets (Deutsch and Journel, 1998) to smooth out erratic fluctuations in the computed weights. Choosing an appropriate cell size requires visualizing a plot of the declustered mean to the cell size. Since data are usually preferentially sampled in the high-valued regions, choosing a minimum mean will be more representative than the naïve mean.

Kriging declustering takes advantage of the property of kriging weights to account for the redundancy of information in the sample set due to geometrical configuration of the data (see Problem 5.1 for further explanation). Kriging weights are the solution to an optimization that rewards closeness and penalizes spatial redundancy. The nuances of method are actually quite complex since the behavior of kriging weights is a complex subject. Kriging declustering may not perform well for some data configurations due to issues, such as data screening (see Problem 5.3 and Deutsch, 2002; Chilès and Delfiner, 1999; Goovaerts, 1997) and boundary effects. In addition, the weights are strongly affected by many of the kriging algorithm parameters, such as search neighborhood and the parameters of the variogram model.

Inverse distance weighting (IDW) declustering simply uses the weights obtained from some inverse distance criterion. The general approach is to assign the weights based on the sum of the inverse distance values for n closest neighbors:

$$w_i^{(id)} = \frac{d_i^{-c}}{\sum_i^{nclose} d_i^{-c}} \tag{3.17}$$

Where the superscript *id* denotes inverse distance, *nclose* is the number of closest neighbors, and c is an exponent selected as the distance criterion in Equation (3.17). In practice, c lies between 1 and 3. The higher the value of c, the greater the weight assigned to nearby data.

Kriging and IDW declustering sum the weights assigned to each datum location during the process of estimating over the unsampled locations in the domain. The cumulative weights are then standardized by the sum of the weights and multiplied by the number of data, as with the other declustering methods. With such standardized weights, a weight > 1 implies the datum is being over weighted and vice versa.

Problem

Compute declustering weights for the dataset *3.3_decluster.dat* (Figure 3.5) using the four methods outlined above. Compare the weights and the declustered univariate statistics between the four methods, rationalize the differences, and select a preferred alternative.

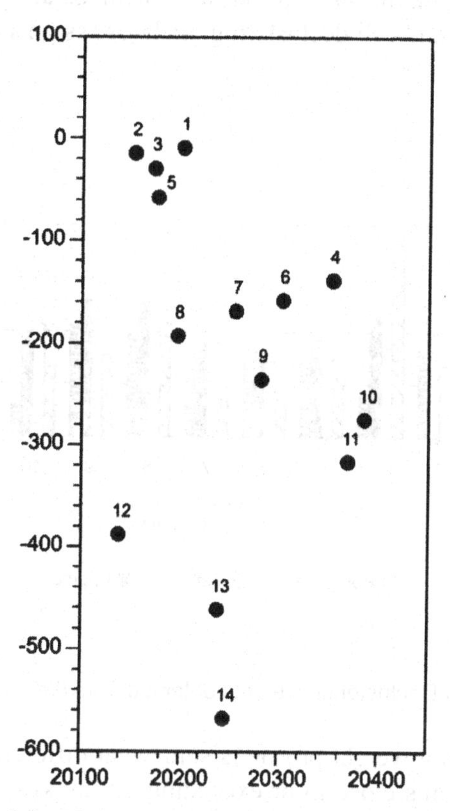

Figure 3.5 Locations of the data in *3.3_decluster.dat* with datum identification number posted for reference.

Solution Plan

Three programs are included to complete this problem, but as usual, the reader is encouraged to write or use their own programs. Program *polydec* performs polygonal declustering; *declus* is a standard GSLIB program to perform cell declustering; and *kt3d_dc* is a modified version of the general GSLIB kriging program. As part of the problem, select or write an inverse distance interpolation program to obtain the inverse distance declustering weights.

For cell declustering, look for the minimum declustered mean since the data are clustered in relatively high-valued areas.

When applying kriging declustering, test the sensitivity of the weights to the type of kriging (simple, ordinary, etc.), and the variogram model, specifically, the range(s) and nugget contribution.

Solution

The declustering weights for each of the four methods are plotted in Figure 3.6. In general, polygonal and cell declustering methods assign similar weights.

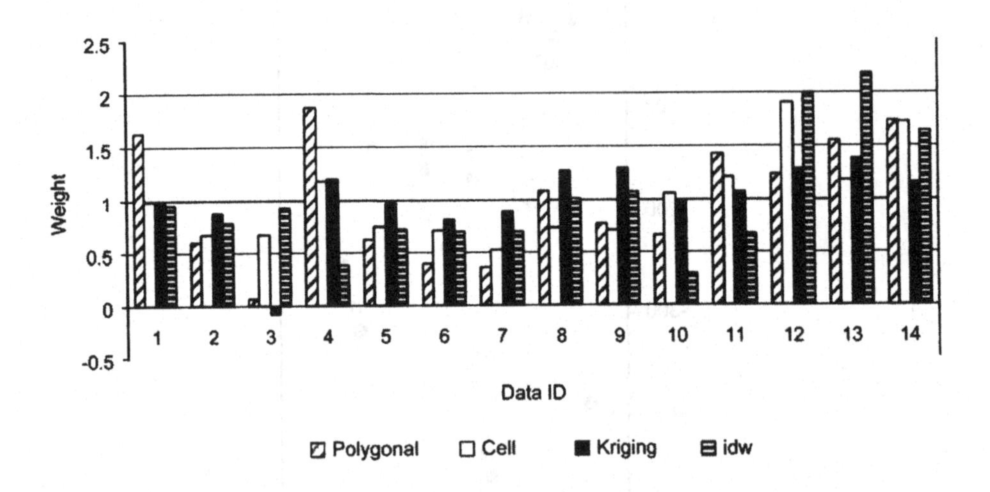

Figure 3.6 Declustering weights obtained from the four methods.

The simple kriging weight profile is somewhat different. Data locations 5-9 are all assigned weights close to or exceeding 1.0 by kriging declustering. The relative magnitude of the kriging weights is somewhat sensitive to the variogram range. The range for this untransformed data set is ~100 m. Significantly over specifying the range, say a 200–300% increase, results in declustering weights that are closer to cell and polygonal declustering weights. It is reasonable to over specify the range rather than to use the modeled variogram range, since declustering is a global debiasing procedure. A global kriging usually refers to estimating with the entire dataset by using a large enough search, but unless the variogram range is long enough to allow far away data to influence the weights, increasing the search alone does not change anything.

The negative weighting of location 3 is due to screening of this data location in estimating most blocks in the domain. It is incidental that polygonal declustering almost equally underweighted data location 3; the data configuration resulted in a very small polygon, so in this case, it worked well.

Cell declustering finds the minimum declustered mean at a cell size ~200 m. Note that this cell size has nothing to do with the desired or optimal model discretization. It simply implies that, due to the data configuration, the most representative (univariate) statistics are obtained when the data are grouped together at this cell size. Cell declustering yields a more uniform set of declustering weights while IDW declustering weights seem to be something of a compromise between kriging and cell declustering.

All methods assign data locations 12, 13, and 14 relatively large weights. This result is expected because they are more widely spaced than the other data.

Remarks

None of these declustering methods consider the data values, but only the geometrical configuration of the data. An inherent assumption and associated limitation follows from this: Traditional declustering methods must assume that the entire range of the distribution has been sampled. Geometrical declustering approaches are not useful if this is not the case. In applying declustering, the most useful results are obtained when the structure of the underlying attribute is understood to some meaningful degree. For example, increasing the amount of nugget contribution in a variogram model, which amounts to understanding less about the underlying spatial structure, results in a much more uniform set of declustering weights. Declustering is of little value outside of a context of spatial correlation.

Although the cell size considered in cell declustering has nothing to do with model discretization considerations, the cell size used in computing kriging declustering weights and IDW weights does matter. Since these latter two are estimation procedures, model discretization impacts the quality of the results.

Boundary effects are pronounced for kriging and polygonal clustering, so it is important to consider the size of the "buffer space" around your data when implementing declustering.

In the end, there are tradeoffs between different methods. Averaging standardized weights from several approaches is an option that may be appropriate in some cases. Note that the approaches in this Problem are not the only available methods to perform declustering. Other methods proposed for declustering involve different statistics: entropy optimized weights (Schofield, 1993), data redundancy measure via the ratio of determinant of correlation matrices with and without the sample (Bourgault, 1997), and an optimized indicator-based definition of the cdf (Bogaert, 1999).

CHAPTER 4

Monte Carlo Simulation

Monte Carlo simulation (MCS) is an algorithm used throughout science and engineering to transfer uncertainty from input variables to output response variables. A realization of all input variables is sampled from their probability distributions. The input probability distributions must be established in a previous step. The realization is processed through a transfer function (TF) to yield output variables. The TF is a numerical model of a process or calculation of interest. This procedure is repeated many times to assemble probability distributions of the output variables. This finding is illustrated in the schematic sketch below.

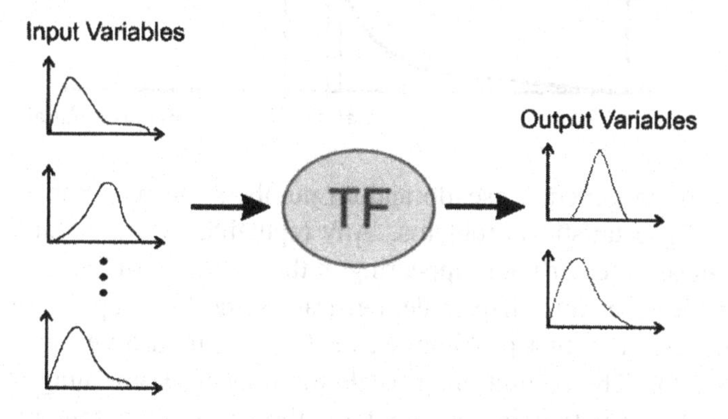

The number of input variables is often large. It may be tens of millions in the context of a 3D geostatistical model. The number of output variables is often much less. If the number of input variables is small, say less than three or four, then we could discretize the multivariate space of the input variables and process

Solved Problems in Geostatistics. By O. Leuangthong, K.D. Khan, and C.V. Deutsch

every possibility; however, the number of possibilities to process increases exponentially with added dimensions. A minimum of 10 values would be needed for each dimension. This leads to a total of 10^N possible values to evaluate for an N-dimensional space. A 10 million cell 3D model would therefore require evaluation of $10^{10,000,000}$ values to fully explore the multivariate space. This number is large with no frame of reference. It is estimated that there are 10^{80} protons in the universe; exploring the full N-D space of input variables is inconceivable when N exceeds 3 or 4. Bellman (1961) referred to this problem as the *curse of dimensionality*, which is where MCS is used. A relatively few realizations (often 20–500) are processed to give a reasonable approximation of the uncertainty in output variables.

The process of drawing samples from a cdf is known as simulation. There are different procedures to simulate from a cdf, but the most flexible and commonly implemented procedure in geostatistics is based on a quantile transformation that consists of two steps: (1) generating a random probability, p, from a uniform distribution [0,1], and (2) finding the quantile value on the cdf corresponding to p. The following schematic illustrates the procedure for three random numbers, $r = 0.805$, 0.435, and 0.344:

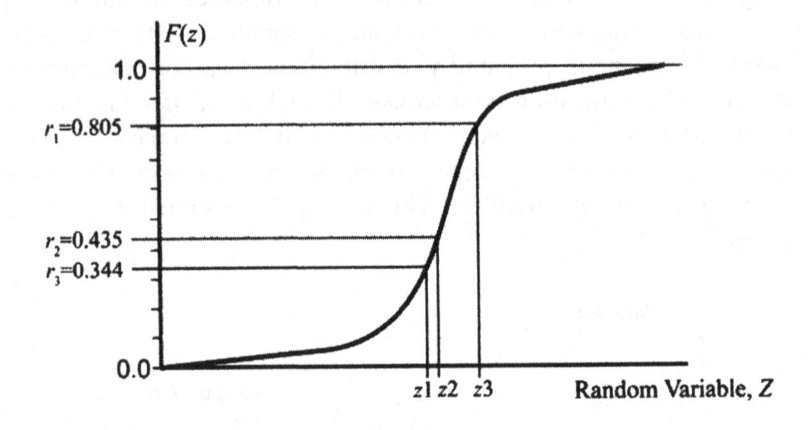

Algorithms to generate pseudorandom numbers are reasonable and suitable for the size of geostatistical problems. Any reputable generator can be used with no risk of unwarranted artifacts appearing in the results of MCS.

The problems in this chapter demonstrate some basic applications of MCS for geostatistics. The first problem uses MCS to demonstrate the Central Limit Theorem (CLT). The second one reveals the impact of assuming independence despite actual correlation in the input on the resulting output distribution of uncertainty. The last problem considers how uncertainty in multiple input variables transfers to uncertainty in a response variable.

4.1 IMPACT OF THE CENTRAL LIMIT THEOREM

Learning Objectives

This problem introduces MCS to show that the uncertainty distribution of a linear combination of variables is Gaussian as predicted by the CLT.

Background and Assumptions

The TF in MCS is often nonlinear; however, if it happens to be a linear sum or average, then the shape of the resulting distribution is predicted to be the Gaussian or normal distribution. The CLT states that the sum of independent identically distributed random variables will be approximately normally distributed. The CLT can be demonstrated by MCS. Consider drawing *N* samples from any arbitrary distribution of finite variance. The sum or average of the *N* samples over many realizations will tend to a normal or Gaussian distribution.

The CLT states that any sum of a large number of independent and identically-distributed random variables tends toward a normal distribution (Spiegel, 1985, pp. 112–113). For example, consider the average of n RVs, X_i, $i=1,...,n$, where X_i has a mean and variance of μ and σ^2, respectively.

$$\bar{X} = \frac{1}{n}\sum_{i=1}^{n} X_i$$

Based on the CLT, the distribution of $\bar{X}$ will be Gaussian. Simple arithmetic can be used to establish that the mean of the mean will be equal to μ. The variance can be calculated as the following since the X RVs are independent:

$$Var\{\bar{X}\} = \frac{\sigma^2}{n}$$

Note that this result was developed in Problem 2.2. The RVs, X_i, $i=1,...,n$, can follow any distribution; this problem makes use of a uniform distribution to demonstrate the veracity of the CLT.

Problem

(i) Verify that the sum of independent random variables tends toward a normal distribution as predicted by the CLT. For this problem, consider 10 random variables, X_i, $i=1,\ldots,n$, with a uniform probability distribution between 0 and 1, and create a RV, S, that is the sum of these 10 uniform RVs.

$$S = \sum_{i=1}^{10} X_i \tag{4.1}$$

Generate several realizations of S, and calculate its mean and variance.

(ii) What is the mean and variance of the probability distribution uniform between 0 and 1? Verify that the mean and variance of the sum of 10 uniform random variables in Part (i) is correctly predicted by the CLT.

Solution Plan

(i) Set up a grid of 100 rows by 10 columns in a spreadsheet with uniform random numbers between 0 and 1. Then create an eleventh column with the sum of these first 10 columns. Calculate the average and variance of this eleventh column. Note that precoded spreadsheet functions may return the "sample variance", which has a slightly different formula than the variance. Consider coding your own formula for the variance, which is expressed as:

$$\sigma_S^2 = \frac{1}{n}\sum_{i=1}^{n}\left(S_i - \bar{S}\right)^2$$

The use of $1/n$ instead of $1/(n-1)$ leads to a biased estimate of the variance in presence of a small sample. The assumption in the use of $1/n$ is that the sample of n data can be taken as the population.

(ii) The mean and variance of the uniform distribution can be found in any statistical reference. Alternatively, you can work it out by plotting the pdf of the uniform distribution for 0-1, and calculating the moments analytically (similar to Problem 2.1). The CLT states that the mean of 10 values summed together should be the mean of the uniform RVs multiplied by 10: check against Part (i). Simple arithmetic tells us that variance is multiplied by 10: again, check against Part (i).

Solution

(i) The average and variance of the sum, S, of the 10 uniform random numbers is ~5.0 and 0.83. The exact values of these summary statistics depends on the random number generator, so variations from these values are expected. Further, with many realizations of S (i.e., >100), we should expect that the calculated statistics should be even closer to 5.0 and 0.83, respectively for the mean and variance. Figure 4.1 shows one example of the histogram of S and the corresponding summary statistics.

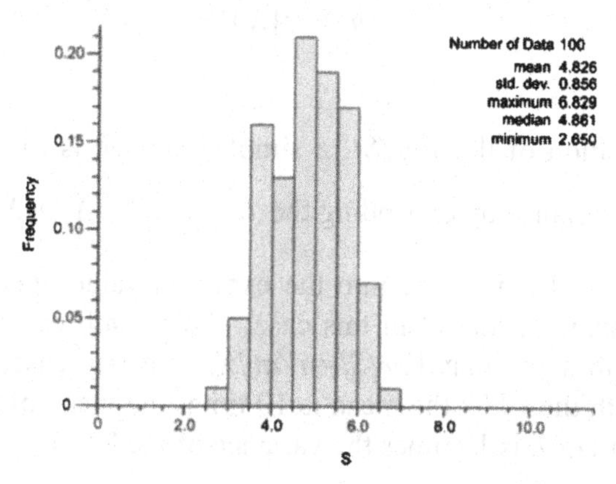

Figure 4.1 Example histogram of S, the sum of 10 uniform RVs.

(ii) The mean and variance of the uniform distribution is

$$m = \frac{a+b}{2}$$

$$\sigma^2 = \frac{1}{12}(b-a)^2$$

Where a and b are the limits of the uniform distribution, that is, the RV is uniformly distributed between a and b. In this case, $a=0$ and $b=1$, so the mean and variance of the uniform distribution is 0.50 and 0.083, respectively.

Given Equation (4.1), we could derive the mean and variance of the sum by using the expected value operator (only partial steps are shown below):

$$E\{S\} = E\left\{\sum_{i=1}^{n} X_i\right\}$$
$$= \sum_{i=1}^{n} E\{X_i\}$$
$$= n \cdot E\{X\}$$

$$Var\{S\} = E\{S^2\} - [E\{S\}]^2$$
$$= n \cdot Var\{X\}$$

For the derivation of the *Var{S}*, a simple exercise is to consider *n=2* and solve for the variance by expanding the $S = \sum_{i=1}^{2} X_i = X_1 + X_2$. Substitute this expanded form of *S* for *n=2* into the expected value operations above and show that *Var{S}=2Var{X}* in this case. This result can then be extend to any value *n* to show that *Var{S}=nVar{X}*. In any case, these results are consistent with the CLT: the mean is 10 times the mean of the uniform RVs, X_i; and the variance is 10 times the variance of the RV, X_i.

Remarks

This problem is simple and yet significant. The CLT is pervasive. We commonly consider a sum of RVs, which subtly invokes the normal distribution. The parameters of the resultant normal distribution are predictable. This final point is particularly relevant in geostatistics, since conventional estimation techniques rely on a linear regression approach for inference of the parameters of conditional probability distributions.

Note that the results in this problem are similar to, but not exactly like, those developed in Problem 2.2. In that case, we considered the distribution of the averages while only the sum is considered in this problem. As a result, the variance of this linear combination is different than the one evaluated in Problem 2.2. In most practical cases, however, we are interested in the average. As seen in this problem and Problem 2.2, the shape of the distribution of uncertainty is predicted by the CLT and the moments are estimated by linear algebra and expected values. There is no need to use MCS to establish the distribution of uncertainty in the average; however, the problem demonstrates MCS and the CLT.

Data availability was not an issue in this case; however, in practical applications, data sparsity is a real challenge. The next section focuses on an interesting application of MCS to assess the uncertainty in a summary statistic when we are confronted with this challenge.

4.2 BOOTSTRAP AND SPATIAL BOOTSTRAP

Learning Objectives

The objective of this problem is to understand the principles and application of MCS used in the bootstrap. The bootstrap is a resampling method that can be used to evaluate uncertainty in sample statistics. In the first part of this problem, the traditional bootstrap is employed, which assumes independence between the data. In the second part, we will evaluate how spatial correlation in the data can cause the uncertainty to be larger than if they are assumed independent. This finding is due to a reduction in the number of effectively independent data.

Background and Assumptions

The bootstrap method allows the assessment of uncertainty in a statistic by using the actual data (Efron, 1982; Efron and Tibshirani, 1993). The bootstrap procedure is performed as follows:

1. Assemble the representative histogram of the n data, using declustering and debiasing techniques if appropriate, to obtain the distribution of the Z random variable: $F_Z(z)$.
2. Draw n values from the representative distribution, that is, generate n uniformly distributed random numbers p_i, $i=1,\ldots,n$ and read the corresponding quantiles: $z_i = F_Z^{-1}(p_i)$, $i=1,\ldots,n$. The number of data drawn is equal to the number of data available in the first place. The distribution of simulated values is not identical to the initial data distribution because they are drawn randomly and with replacement.
3. Calculate the statistic of interest from the resampled set of data.
4. Return to Step 2 and repeat L times, for example, L=1000.
5. Assemble the distribution of uncertainty in the calculated statistic.

For example, if 10 data are available and we calculate the mean, there is undoubtedly uncertainty about the mean. In the bootstrap approach, we sample, with replacement, 10 values from the data set and recalculate the mean. Note that when we sample with replacement, it is possible (though highly unlikely) to draw the same data value all 10 times. By repeating this resampling procedure many times, say 1000 times, we obtain 1000 possible mean values. The distribution of these 1000 sampled means shows the uncertainty in the mean when only 10 samples are available. In this way, we assess uncertainty in the data by using the data themselves. This process is entirely internal, like "pulling oneself up by their bootstraps"; hence the name given to this method. This method is simple, effective and automatic to implement; however, it assumes that the available data are representative of the entire field or population, and that the samples are independent.

The spatial bootstrap (Deutsch, 2004) is an extension of the bootstrap concept to cases where spatial correlation exists. In general, spatial correlation is quantified using a covariance function dependent on spatial coordinates. This function is used to construct a covariance matrix that summarizes the correlation between all available data locations. This covariance matrix is then used to determine the set of drawn values such that the spatial correlation between data values is preserved (on average, over multiple sets of samples). While there is no standard method for performing the spatial bootstrap; a number of methods have been implemented. The program provided for this problem performs a Cholesky (*LU*) decomposition of the n x n covariance matrix, built from the variogram (or covariance) model; the basis of this type of simulation is matrix or LU simlulation (Alabert, 1987; Davis, 1987). Specific details are provided in the supplementary materials in *Spatial_Bootstrap.pdf*.

This problem is designed around a small number of wells for a mixed siltstone–carbonate reservoir in west Texas. The original 3D data is converted into well-averaged attributes to facilitate working in a simpler 2D setting; this data can be found in *4.2_2Dwelldata.dat*.

Problem

(i) Examine the spatial configuration of the data and determine whether declustering is required. If so, then perform declustering and determine the representative mean of the 2D well-averaged porosity. Also, calculate the correlation between the average porosity and the collocated seismic data (using declustering weights if required). Perform the bootstrap to evaluate the uncertainty in these summary statistics. You can write your own program or use the program *boot_avg* and *boot_corr* from the supplementary materials to perform the bootstrap for the average and the correlation coefficient, respectively.

(ii) Now consider that the data are spatially correlated and that this correlation is modeled by a spherical variogram with a range of 8500 m in the N–S direction and 6500 m in the E–W direction with a 10% nugget contribution. Mathematically, this correlation model is written

$$\rho(\mathbf{h}) = \begin{cases} 1.0 & \mathbf{h} = 0 \\ 0.9 \cdot \left[1.0 - 1.5(\mathbf{h}) + 0.5(\mathbf{h})^3\right] & 0 < \mathbf{h} \le 1 \\ 0.0 & \mathbf{h} > 1 \end{cases}$$

where **h** is the Euclidean distance vector between data pairs standardized by the range, a:

$$\mathbf{h} = \sqrt{\left(\frac{h_x}{a_x}\right)^2 + \left(\frac{h_y}{a_y}\right)^2}$$

Here $h_x = x_i - x_j$ and $h_y = y_i - y_j$ for the ith and j^{th} data samples, $i,j=1,\ldots,n$, and a_x and a_y are the range of correlation in the x and y directions, respectively. In this case, the range in the N–S or y-direction is 8500 m (i.e., a_y=8500), and the range in the E–W or x-direction is 6500 m (i.e., a_x=6500). Given this correlation between the data, perform a spatial bootstrap to re-evaluate the uncertainty in the mean porosity. Plot the distribution of uncertainty and note the variance of the mean. Compare these results to the assumption of independence in Part (i). The program *spatial_bootstrap* is available to perform this task.

Solution Plan

(i) While the data are fairly regularly sampled over the field, it is good practice to determine if declustering will impact the statistics. So the first task is to determine the representative distribution on which we can then perform the bootstrap. Cell declustering can be used in this case (although other declustering approaches could also be used). If there is an appreciable difference in the declustered mean, then consider using the declustering weights to construct the representative distribution. Use the *boot_avg* program to perform a bootstrap of the mean.

The correlation of the average porosity and the collocated seismic can be calculated using the covariance from Equation (2.12) and then standardizing this covariance by the standard deviation of the average porosity and the standard deviation of the seismic data. Note that in all instances, these calculations should use declustering weights if declustering is deemed necessary. Use the program *boot_corr* to bootstrap the correlation coefficient.

(ii) Run the *spatial_boostrap* program and note the variance of the mean. We could also consider calculating the number of effective or independent data. The variance of the mean reduces by *1/n* if the data are independent; therefore, the effective number of data can be calculated as:

$$\sigma_{\bar{z}}^2 = \frac{\sigma_z^2}{n_{eff}} \quad \text{or} \quad n_{eff} = \frac{\sigma_z^2}{\sigma_{\bar{z}}^2} \tag{4.2}$$

where σ_z^2 is the variance of the sample data and $\sigma_{\bar{z}}^2$ is the variance of the sample means.

Solution

(i) The histogram of the original and declustered porosity data and the cross plot of porosity with the collocated seismic data are shown in Figure 4.2. Cell declustering was performed using the program *declus* (from Problem 3.3). The declustered distribution shows a slightly lower mean for porosity of 8.17% compared with the 8.40% given by the equal-weighted distribution. Declustering also impacts the correlation coefficient by reducing this value from 0.62 to 0.60. The following results from bootstrapping the mean and correlation are based on using the declustered distribution.

The parameters used in the program, *boot_avg*, for resampling of the well data to evaluate the uncertainty in the sample mean are as follows:

```
                 START OF PARAMETERS:
Line 1.  declus.out             -input file with histogram
Line 2.  4   8                  -column for value and prob
Line 3.  boot_avg.out           -output file
Line 4.  62  1000               -ndraw, nsim
Line 5.  69069                  -Random Number Seed
```

Note that Line 1 calls on the output file from declustering, not the original *4.2_2Dwelldata.dat*. The output file from declustering contains the original data in *4.2_2Dwelldata.dat* with the inclusion of an extra column for the declustering weights. The parameters in Line 4 indicate that 62 samples will be drawn during each of 1000 resampling trials to be performed. Using these parameters, the resultant distribution of sample means is shown in Figure 4.3 (a).

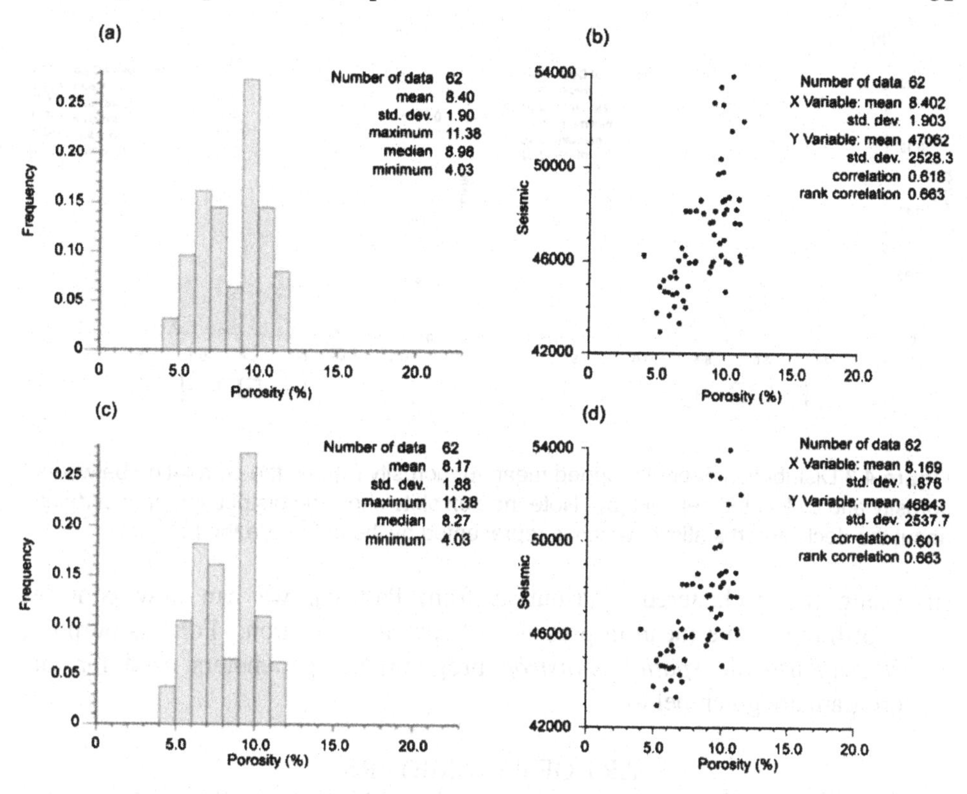

Figure 4.2 Original porosity data (a) and cross plot with seismic data (b), and the declustered distribution of porosity (c) and declustered cross plot (d).

The program *boot_corr* can be used to bootstrap the correlation coefficient. The parameters used for this program are as follows:

START OF PARAMETERS:

Line 1.	declus.out	-input file with histogram
Line 2.	4 6 8	-column for xvar,yvar and weight
Line 3.	bootcorr.out	-output file
Line 4.	62 1000	-num, nset
Line 5.	69069	-Random Number Seed

Once again, the input data file is the output file from cell declustering that contains the declustering weights. The resultant distribution of bootstrapped correlation coefficients is shown in Figure 4.3 (b).

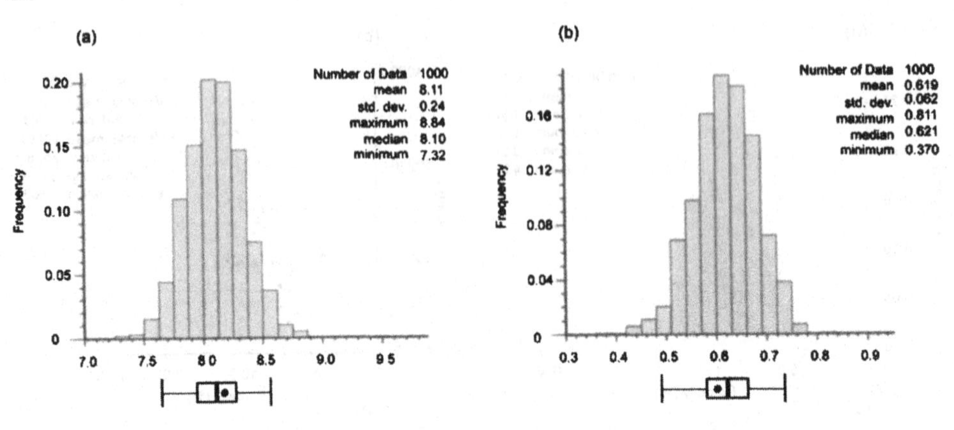

Figure 4.3 Distribution of bootstrapped mean for porosity (a) and the correlation between porosity and collocated seismic (b). Note the box plots show the distribution of uncertainty about the declustered statistic, which is represented by the dot inside the box plot.

(ii) Using the declustered distribution from Part (i), we can now consider bootstrapping the mean in presence of spatial correlation. For this purpose, we can use the *spatial_bootstrap* program; the parameters used for this program are given below:

```
                        START OF PARAMETERS:
Line1.  declus.out                      -file with reference distribution
Line2.  4 0                             -  columns for value and weight
Line3.  −1.0e21  1.0e21                 -trimming limits
Line4.  4.2_2DWellData.dat              -file with locations to resample
Line5.  2 3 0                           -  columns for X, Y, Z locations
Line6.  0                               -save realizations? (0=no, 1=yes)
Line7.  SB-realizations.dat             -file for all simulated realizations
Line8.  SB-mean.dat                     -file for bootstrapped average
Line9.  0.0                             -threshold (average above is reported)
Line10. 1000                            -number of realizations
Line11. 69069                           -random number seed
Line12. 1  0.0                          -nst, nugget effect
Line13. 1  1.0 0.0 0.0 0.0              -it,cc,ang1,ang2,ang3
Line14.    8500.0 6500.0 10.0           -a_hmax, a_hmin, a_vert
```

Unlike the previous two programs where the parameters are reasonably self-explanatory, the *spatial_bootstrap* program requires additional parameters that are not as obvious. Line 5 indicates the columns in the data file with the spatial coordinates of the sample locations. Line 6 gives the option to save every drawn value, such that if we choose to simulate 5000 sets of data (as currently set in Line 10), then all 1000 realizations of 62 drawn values are written to an output file specified in Line 7. Line 8 gives the output file for the 1000 bootstrapped correlation coefficients. While Line 9 permits a threshold to be imposed in the event that a specific range of value is of interest (e.g., economic thresholds on the attribute of interest could be used). Line 10 specifies the number of resampling trials to perform. Lines 12–14 specify the variogram model (i.e., the spatial correlation model between the data) according to GSLIB convention (Deutsch and Journel, 1998).

The distribution of the sample means in the presence of spatial correlation is shown in Figure 4.4 and the effective number of data, n_{eff}, calculated from Equation (4.2) is 15.63. The impact of dependence between the data is more uncertainty in the summary statistic (i.e., greater variability, hence greater spread of values).

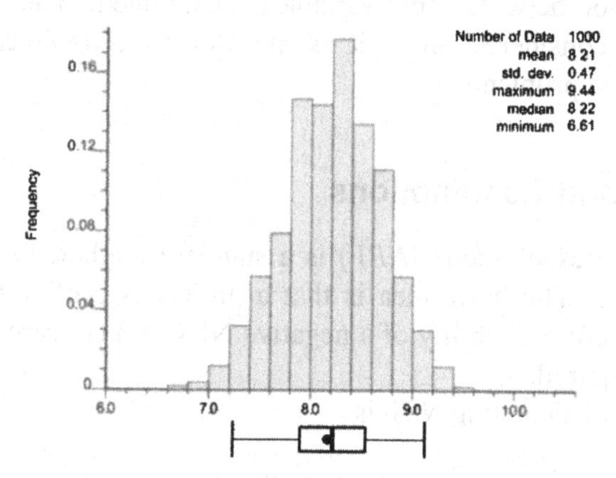

Figure 4.4 Distribution of mean porosity considering spatial correlation between the data. Compare to the distribution of sample means under the independent resampling of the data in Part (i).

Remarks

The bootstrap is typically applied within deemed homogeneous geological rock types to establish uncertainty in parameters, such as the mean, variance, and distribution shape. The uncertainty in the rock types and their proportions must be considered separately.

A key assumption in the bootstrap is that the entire range of the underlying population has been sampled by the data. If this condition is not met, the bootstrap will underestimate uncertainty. In these instances, debiasing should be considered prior to bootstrapping.

The increased variance in the sample mean in the case of spatial correlation between the data is a result of the reduction in the effective number of independent data. Although the relationship of Equation (2.15) shows this from the CLT (see Problems 2.2 and 4.1), the result is somewhat counterintuitive since we might expect that correlation between the data should reduce the uncertainty in a typical Monte Carlo procedure.

4.3 TRANSFER OF UNCERTAINTY

Learning Objectives

This problem shows how uncertainty in a set of input parameters is transferred through a function to yield uncertainty in the sought-after model response. The effect of correlation between input variables is examined. The net present value of a project is considered since it is an important nonlinear TF in many engineering related problems.

Background and Assumptions

Calculating Net Present Value (*NPV*) is a standard method for screening long-term investments. The basic idea is that a project is profitable if the NPV is positive. A nonzero probability of a negative NPV is a concern, and a minimum return may be required.

A formula for calculating *NPV* is

$$NPV = \sum_{i=1}^{n} \frac{R_i - C_i}{(1+r)^i} \tag{4.3}$$

where, R is annual revenue, C is costs, r is the discount rate, and n is the number of years. If the revenue and costs can be modeled as a RV, then Monte Carlo Simulation can be used to quickly evaluate the probability distribution of NPV, given the uncertainty model for these two input parameters.

Problem

(i) Evaluate the uncertainty in *NPV* for the first 5 years of a proposed project given the following parameter distributions provided by a finance model: Annual revenue, R, may be modeled by a normal distribution with mean and variance of \$10 million and 2 million^2, respectively. Annual costs, C, are also assumed to follow a normal distribution with mean and variance of \$8 million and 2 million^2. Choose a constant discount factor, r, say, 10%.

Calculate the P10, P50, and P90 of NPV. What is the probability of a negative NPV?

(ii) Repeat the above evaluation of NPV assuming a correlation coefficient of 0.80 exists between revenue and costs and that the bivariate distribution between these two variables is Gaussian. You should write your own program for these problems. Alternatively, a simulation plug-in for MS Excel may be more convenient.

Solution Plan

(i) The program flow for NPV simulation is as follows:

a) Simulate n random numbers, p_i, $i = 1,\ldots,n$, and obtain corresponding quantiles, $z_i = F_Z^{-1}(p_i)$, $i = 1,\ldots,n$ from the non standard Gaussian distribution for each RV C, and R.

b) Calculate *NPV* as $NPV = \sum_{i=1}^{5} \frac{R_i - C_i}{(1+r)^i}$.

c) Repeat L times to assemble the distribution for *NPV*.

To obtain a reasonably reliable estimate of the P10, P50 and P90, the number of realizations of NPV should be large, say L=1000.

To calculate the P10, P50, and P90, construct the cdf of the simulated NPV by sorting the output realizations in ascending order and assign the corresponding probability to assemble a cdf and read the required quantiles directly from the cdf. Similarly, determine what probability of the simulated NPV is negative from this same cdf.

(ii) Assuming a bivariate normal pdf with $\rho_{RC} = 0.8$, the conditional mean and variance of R are given by Johnson and Wichern (1998, pp. 171–172):

$$E\{R_i \mid C_i = c\} = m_R + \rho_{R,C} \frac{\sigma_R}{\sigma_C}(c - m_C) \tag{4.4}$$

and

$$Var\{R_i \mid C_i = c\} = \sigma_R^2 \left(1 - \rho^2_{R,C}\right) \tag{4.5}$$

Unlike Part (i), where a simulated value was drawn for C and R, this part requires that a random deviate is drawn for C, but the random deviate drawn for R must come from the conditional distribution with parameters specified by Equations (4.4) and (4.5). Thus every realization of R is dependent on the value of C that is drawn.

Solution

(i) The resultant distribution of *NPV* for the case of independence between the input parameters is shown in Figure 4.5. The P10, P50, and P90 *NPV* corresponds to values of ~$3.4, 7.5, and 12.1 million, respectively. The probability of a negative *NPV* is 0.01.

(ii) Considering the specified correlation between R and C, the distribution for *NPV* is shown in Figure 4.6. The P10, P50, and P90 *NPV* corresponds to values of ~$5.8, 7.7, and 9.6 million, respectively. The probability of a negative *NPV* is 0.0.

Remarks

Although input variables may be normally distributed, the output of a transfer function need not necessarily follow a Gaussian distribution. This result is not uncommon; in fact, the only instance in which the distribution of the response would be Gaussian is the case where the response variable is a linear combination of the input variables. For the case of NPV, the consideration of the time value of money makes the calculation non-linear.

In the somewhat unrealistic case of independence between revenues and costs, there is a small probability of a negative *NPV*. In the case of correlation between R and C, there is less uncertainty in *NPV* and a zero probability of negative *NPV*. Thus it is always important to identify and consider correlation in the simulation input variables. However, Monte Carlo simulation often treats the input parameters as independent since the underlying distributions of the input variables, and the joint distributions are not known.

Haldorsen and Damsleth (1990) give a nice spatial context to the use of MCS to evaluate uncertainty in a response variable due to uncertainty in multiple input variables.

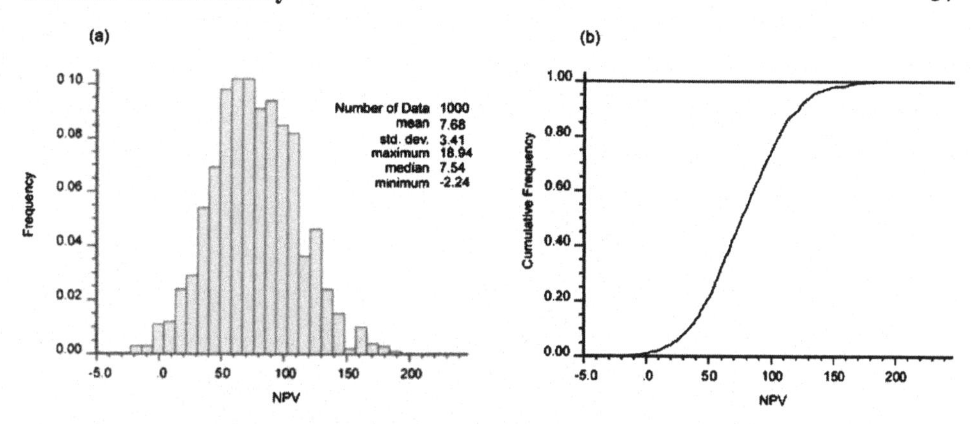

Figure 4.5 Histogram (a) and cumulative histogram (b) of simulated *NPV* assuming independence between the input variables.

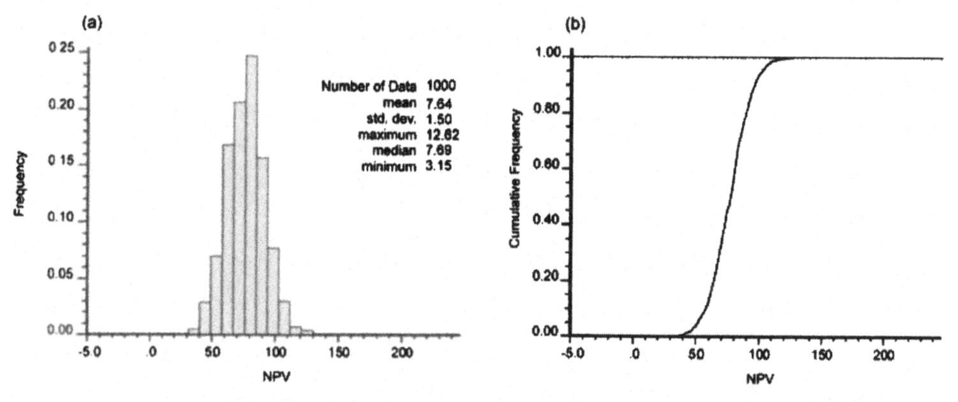

Figure 4.6 Histogram (a) and cumulative histogram (b) of *NPV* with correlated revenue and cost inputs. Note the reduction in variance.

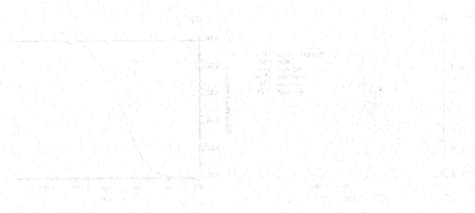

CHAPTER 5

Variograms and Volume Variance

An end objective of geostatistics is quantifying and exploiting spatial patterns of regionalized variables. The most common measures of spatial structure are the covariance and variogram functions, of which the variogram is most commonly measured in practice. The variogram is denoted by $2\gamma(\mathbf{h})$, and is a two-point statistic used to quantify the variability between two random variables separated by a lag vector, **h**:

$$2\gamma(\mathbf{h}) = E\left\{\left[Z(\mathbf{u}) - Z(\mathbf{u}+\mathbf{h})\right]^2\right\}$$

where $Z(\mathbf{u})$ is a spatially regionalized random variable. $\gamma(\mathbf{h})$ is the *semi*variogram or, by common usage, loosely called the variogram. Any time the term "variogram" is used in this book, it is the semivariogram $\gamma(\mathbf{h})$ that is intended.

The stationary variance of $Z(\mathbf{u})$ and $Z(\mathbf{u}+\mathbf{h})$ could be written as $\sigma^2 = C(\mathbf{h}=0) = E\{Z^2\} - m^2$. The spatial covariance between $Z(\mathbf{u})$ and $Z(\mathbf{u}+\mathbf{h})$ would be written, assuming the mean, m, is the same at locations **u** and **u**+**h** (a stationary mean):

$$C(\mathbf{h}) = E\left\{\left[Z(\mathbf{u}) - m\right]\left[Z(\mathbf{u}+\mathbf{h}) - m\right]\right\} = E\left\{Z(\mathbf{u}) \cdot Z(\mathbf{u}+\mathbf{h})\right\} - m^2$$

Solved Problems in Geostatistics. By O. Leuangthong, K.D. Khan, and C.V. Deutsch

Expanding the expression for the variogram assuming a stationary mean and the relations recalled above:

$$\begin{aligned} 2\gamma(\mathbf{h}) &= E\{Z(\mathbf{u})^2\} - m^2 - 2E\{Z(\mathbf{u})\cdot Z(\mathbf{u}+\mathbf{h})\} \\ &\quad +2m^2 + E\{Z(\mathbf{u}+\mathbf{h})^2\} - m^2 \\ \gamma(\mathbf{h}) &= \sigma^2 - C(\mathbf{h}) \quad \text{or} \quad C(\mathbf{h}) = \sigma^2 - \gamma(\mathbf{h}) \end{aligned} \tag{5.1}$$

These relations are valid for a regionalized variable with a stationary mean and variance.

The problems in this chapter focus on calculating and modeling the variogram. The first problem develops an understanding of geometric anisotropy; or the phenomenon of directional dependence for the continuity of spatially distributed variables and emphasizes that it is often apparent variogram ranges we work with. The second problem illustrates how to compute the variogram using experimental data. The last problem introduces fitting a variogram model to experimentally calculated values, and the impact of choosing a particular volume support.

5.1 GEOMETRIC ANISOTROPY

Learning Objectives

This problem covers the operations involved in rotating a local coordinate system in order to align the principal axes with the principal directions of geologic continuity. This is known as specifying a model with geometric anisotropy.

Background and Assumptions

Regionalized variables are rarely truly isotropic. There are almost always directions of greater and lesser continuity. Anisotropy requires a specification of the orientation of the coordinate system and the magnitude of the anisotropy. The orientation of the principal directions of continuity is specified with one rotation angle in 2D and three angles in 3D. The magnitude of anisotropy is specified by two range parameters in 2D and three range parameters in 3D.

Consider two locations separated by a vector **h**. The distance between these points must be calculated accounting for the anisotropy. The vector **h** may be defined by the separation distances in the three original *x*, *y*, and *z* Cartesian coordinate systems: Δx, Δy, and Δz. These separation distances are rotated into the coordinates aligned with geological continuity: Δx_R, Δy_R, and Δz_R. The procedure for this rotation will be reviewed below. A dimensionless distance is calculated knowing the ranges of correlation in the rotated coordinate system:

$$h = \sqrt{\left(\frac{\Delta x_R}{a_1}\right)^2 + \left(\frac{\Delta y_R}{a_2}\right)^2 + \left(\frac{\Delta z_R}{a_3}\right)^2}$$

This dimensionless distance, h, is 0.0 when the two locations are coincident, 1.0 when the distance between the two locations is equal to the range, a, and > 1.0 when the two locations are separated by a distance greater than the range. Note that the range follows the surface of an ellipsoid with principal axis distances a_1, a_2, and a_3. This was introduced briefly in Problem 4.2 with the spatial bootstrap.

Variogram models are 1D functions of the dimensionless scalar distance h that convert this Euclidean distance to a variogram measure of 'geological distance'. Relatively few variogram models are used in common practice. These models constitute a physically valid measure of distance (Christakos, 1984; Goovaerts, 1997, p. 87). One such model, for illustration, is the spherical model

$$\gamma(h) = c \cdot Sph(h) = c \cdot \begin{cases} \left[1.5h - 0.5h^3\right], & \text{if } h \le 1 \\ 1, & \text{if } h \ge 1 \end{cases}$$

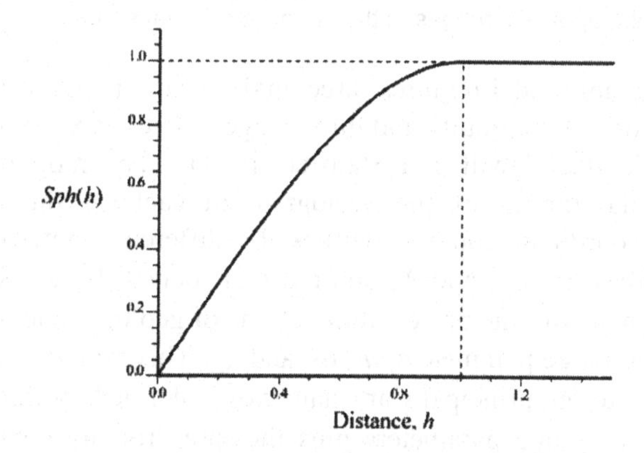

where c is the variance scaling parameter. The range is normalized to 1 because the range parameters a_x, a_y, and a_z were used in the calculation of the dimensionless distance, h.

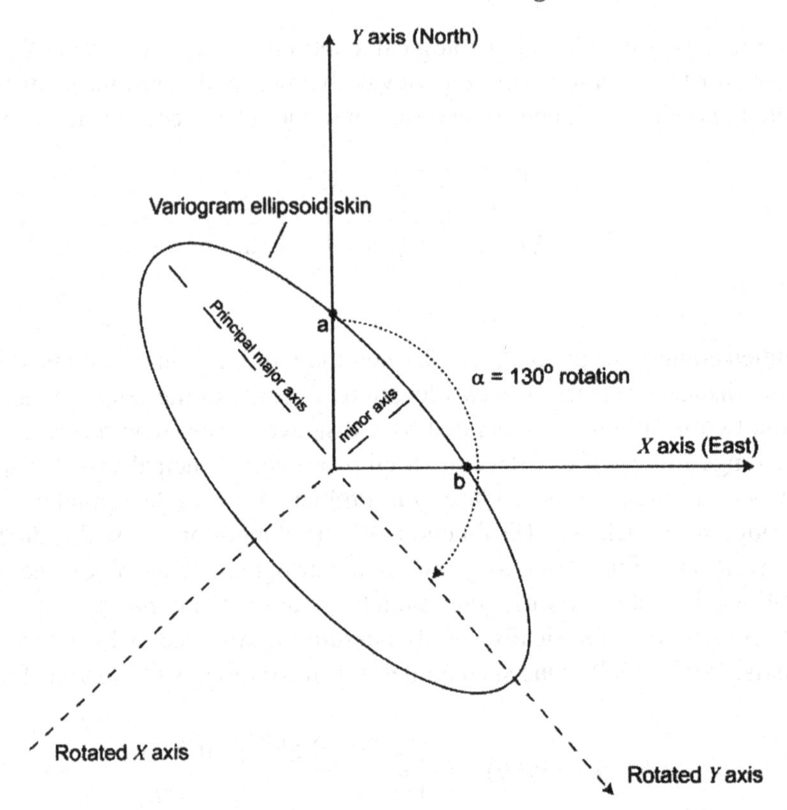

Figure 5.1 Depiction of rotation of a variogram ellipsoid in the plane containing the *X* and *Y* axes of a Cartesian coordinate system, with the *Z*-axis perpendicular to the page. The points (a) and (b) mark the distances along the principal axes of the coordinate system corresponding to the apparent ranges of the variogram in that space.

A 3D variogram model requires three angles to define the orientation of the principle directions of continuity and three ranges. These six parameters define a tri-axial ellipsoid (Gendzwill and Stauffer, 1981). The radii of the ellipsoid correspond to the ranges of the variogram in each of the three principal directions. A coordinate rotation defines the difference between the original unrotated distances Δx, Δy, and Δz and a set of rotated distances Δx_R, Δy_R, and Δz_R that are aligned with the principal directions of geologic continuity called for by the variogram range parameters, a_1, a_2, and a_3. This rotation of the variogram ellipsoid relative to the principal Cartesian axes is defined by three angles, α, β, and θ. The three range parameters plus the three rotation angles completely specify the geometry of a variogram model.

The specification of the rotation angles is a potential source of confusion since there are a number of possible ways to define them. In this book, the conventions used in Deutsch and Journel (1998) are adopted. The principal major axis (the maximum range of the variogram, a_1) is oriented along the rotated *Y*-direction in the rotated or non rotated coordinate system (Figure 5.1). The principal minor axis (the minimum 'horizontal' range of the variogram, a_2) is oriented along the rotated X-direction. The perpendicular axis (the 'vertical' range of the variogram, a_3) is oriented along the rotated Z-direction.

Considering a non rotated 3D Cartesian coordinate system, with X and Y axes in the horizontal plane and Z-axis oriented vertically, the rotation angles are defined as follows:

Angle 1 (α) is obtained by rotating the original X- and Y-axes (in the horizontal plane) about the original vertical Z-axis. This angle is measured between the Y-axis, having an azimuth of 0.0° (north) and the rotated Y-axis, in positive degrees clockwise looking downward (in the direction of decreasing z) toward the origin. The new intermediate coordinate system is denoted X_I, Y_I, Z. This step is an azimuth correction since the coordinate system is rotated in the horizontal plane. The new 'intermediate' rotated coordinates, X_I and Y_I, are given by the following equations:

$$
\begin{aligned}
X_I &= x\cos\alpha + y\sin\alpha \\
Y_I &= x\sin\alpha - y\cos\alpha \\
Z &= Z
\end{aligned}
$$

and are shown schematically in Figure 5.2.

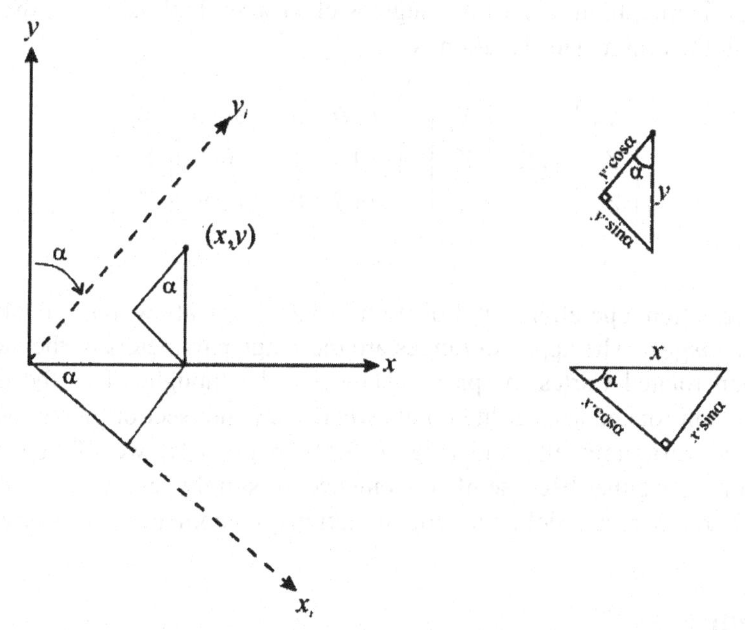

Figure 5.2 Schematic illustration of first rotation of azimuth angle, α. This rotation happens about the Z-axis. In this instance, the solution is for (x_i, y_i) given (x,y).

Equivalently, we can consider the matrix notation for the above system of equations:

$$\begin{bmatrix} X_I \\ Y_I \\ Z \end{bmatrix} = [\mathbf{R}_1] \begin{bmatrix} X \\ Y \\ Z \end{bmatrix} = \begin{bmatrix} \cos\alpha & -\sin\alpha & 0 \\ \sin\alpha & \cos\alpha & 0 \\ 0 & 0 & 1 \end{bmatrix} \begin{bmatrix} X \\ Y \\ Z \end{bmatrix} \tag{5.2}$$

Angle 2 (β) is defined via the rotation of the preceding configuration about the X_I-axis such that the Y_I-axis is directed down a specified dip direction. The new orientation of the Y_I-axis is denoted Y_R, and is the final rotated orientation of the Y-axis. This angle is measured in positive degrees, counter clockwise about the X_I-axis looking toward the origin along the X_I-axis. Similar to the above rotation for angle α, the system of equations for this rotation is obtained as:

$$\begin{bmatrix} X_I \\ Y_R \\ Z_I \end{bmatrix} = [\mathbf{R}_2] \begin{bmatrix} X_I \\ Y_I \\ Z \end{bmatrix} = \begin{bmatrix} 1 & 0 & 0 \\ 0 & \cos\beta & \sin\beta \\ 0 & -\sin\beta & \cos\beta \end{bmatrix} \begin{bmatrix} X_I \\ Y_I \\ Z \end{bmatrix} \tag{5.3}$$

Angle 3 (θ) is obtained by rotating the coordinate system about the Y_R-axis. This angle is measured in positive degrees clockwise, looking along the Y_R-axis away from the origin. The equation is

$$\begin{bmatrix} X_R \\ Y_R \\ Z_R \end{bmatrix} = [\mathbf{R}_3] \begin{bmatrix} X_I \\ Y_R \\ Z_I \end{bmatrix} = \begin{bmatrix} \cos\theta & 0 & -\sin\theta \\ 0 & 1 & 0 \\ \sin\theta & 0 & \cos\theta \end{bmatrix} \begin{bmatrix} X_I \\ Y_R \\ Z_I \end{bmatrix} \tag{5.4}$$

The rotation operations in Equations (5.2)–(5.4) allow the calculation of *apparent ranges*. The apparent ranges are the range parameters of the variogram in the non rotated Cartesian space. They may be thought of as the distances along the non rotated axes at the points where they intersect or pierce the skin of the rotated variogram ellipsoid (Figure 5.1). In practice, we often work with apparent ranges either because of convenience, or simply because it is difficult to accurately measure and define the true directions of geometric anisotropy.

Problem

A 3D regionalized variable has geometric anisotropy with ranges of 150, 90, and 30 m along the principal directions of continuity. These principal directions are rotated from a Cartesian modeling space by angular values of 30°, 20°, and 0°, corresponding to angle 1, angle 2, and angle 3, respectively. What are the apparent ranges in the modeling space?

Solution Plan

(i) Combine the rotation operations shown in Equations (5.2)–(5.4) to give a single rotation matrix, [**R**].

(ii) Multiply the complete rotation matrix, [**R**], by a scaling matrix, [**A**], which normalizes the variogram range parameters in each of the three directions to a value of 1. This step results in an isotropic transformation matrix, [**T**], which will simplify the calculation of the apparent range in the non rotated coordinate system.

(iii) Compute the range components in the three directions oriented parallel to the axes of a non rotated Cartesian coordinate system to obtain the apparent range parameters.

Solution

(i) It is straightforward to combine the three equations (5.2)–(5.4) into a single rotation matrix, [**R**]:

$$\begin{bmatrix} X_R \\ Y_R \\ Z_R \end{bmatrix} = [\mathbf{R}_3] \begin{bmatrix} X_I \\ Y_R \\ Z_I \end{bmatrix}$$

$$= [\mathbf{R}_3][\mathbf{R}_2] \begin{bmatrix} X_I \\ Y_I \\ Z \end{bmatrix}$$

$$= [\mathbf{R}_3][\mathbf{R}_2][\mathbf{R}_1] \begin{bmatrix} X \\ Y \\ Z \end{bmatrix}$$

$$= [\mathbf{R}] \begin{bmatrix} X \\ Y \\ Z \end{bmatrix}$$

where

$$[\mathbf{R}]=[\mathbf{R}_3][\mathbf{R}_2][\mathbf{R}_1]$$
$$=\begin{bmatrix}\cos\theta & 0 & -\sin\theta\\ 0 & 1 & 0\\ \sin\theta & 0 & \cos\theta\end{bmatrix}\begin{bmatrix}1 & 0 & 0\\ 0 & \cos\beta & \sin\beta\\ 0 & -\sin\beta & \cos\beta\end{bmatrix}\begin{bmatrix}\cos\alpha & -\sin\alpha & 0\\ \sin\alpha & \cos\alpha & 0\\ 0 & 0 & 1\end{bmatrix}$$

Since $\theta = 0$, $[\mathbf{R}_3]$ simplifies to an identity matrix, [**I**], and [**R**] simplifies to

$$[\mathbf{R}]=[\mathbf{R}_2][\mathbf{R}_1]$$
$$=\begin{bmatrix}1 & 0 & 0\\ 0 & \cos\beta & \sin\beta\\ 0 & -\sin\beta & \cos\beta\end{bmatrix}\begin{bmatrix}\cos\alpha & -\sin\alpha & 0\\ \sin\alpha & \cos\alpha & 0\\ 0 & 0 & 1\end{bmatrix}$$
$$=\begin{bmatrix}\cos\alpha & -\sin\alpha & 0\\ \sin\alpha\cos\beta & \cos\alpha\cos\beta & \sin\beta\\ -\sin\alpha\sin\beta & -\cos\alpha\sin\beta & \cos\beta\end{bmatrix}$$

(ii) Now with the rotation matrix, [**R**], determine a scaling matrix, [**A**], that normalizes the variogram ranges to unity. This simply requires a diagonal matrix that scales each distance component in the three principal directions by their respective ranges:

$$[\mathbf{A}]=\begin{bmatrix}1/a_2 & 0 & 0\\ 0 & 1/a_1 & 0\\ 0 & 0 & 1/a_3\end{bmatrix}$$

where a_1 and a_2 are the range parameters in the rotated maximum and minimum horizontal directions, and a_3 is the range in the rotated vertical direction. Recall that by the convention adopted here, a_1 is the range in the rotated *Y*-axis direction and a_2 corresponds to the range in the rotated *X*-axis direction. Recall that the goal is to transform distance vectors from one coordinate system to a rotation of that system. For this, we define a transform matrix, [**T**], that accounts for the distance scaling and the rotation:

$$[\mathbf{T}] = [\mathbf{A}][\mathbf{R}^T]$$

$$= \begin{bmatrix} 1/a_2 & 0 & 0 \\ 0 & 1/a_1 & 0 \\ 0 & 0 & 1/a_3 \end{bmatrix} \begin{bmatrix} \cos\alpha & -\sin\alpha & 0 \\ \sin\alpha\cos\beta & \cos\alpha\cos\beta & \sin\beta \\ -\sin\alpha\sin\beta & -\cos\alpha\sin\beta & \cos\beta \end{bmatrix} \quad (5.5)$$

$$= \begin{bmatrix} \cos\alpha / a_2 & -\sin\alpha / a_2 & 0 \\ \sin\alpha\cos\beta / a_1 & \cos\alpha\cos\beta / a_1 & \sin\beta / a_1 \\ -\sin\alpha\sin\beta / a_3 & -\cos\alpha\sin\beta / a_3 & \cos\beta / a_3 \end{bmatrix}$$

(iii) In an isotropic space, the modulus of a vector, $\mathbf{h}$, that reaches the "skin" of the 3D ellipsoid with principal axes describing the variogram ranges in the three principal directions (Figure 5.1) is given by

$$|\mathbf{h}| = \sqrt{(d_1)^2 + (d_2)^2 + (d_3)^2} = 1 \quad (5.6)$$

where: d_1, d_2, and d_3 are distances along the principal axes of a spheroid (i.e., a non rotated, isotropic Cartesian space), such that:

$$\begin{bmatrix} d_1 \\ d_2 \\ d_3 \end{bmatrix} = [\mathbf{T}] \begin{bmatrix} a_x \\ a_y \\ a_z \end{bmatrix} \quad (5.7)$$

with the apparent ranges a_x, a_y, and a_z oriented parallel to the non rotated *X*, *Y*, and *Z* axes, respectively.

In order to solve for the apparent ranges in the non rotated Cartesian system, we must solve for each of a_x, a_y, and a_z separately. Since the apparent range in the non rotated *X*-axis will have coordinates (a_x, 0, 0), then solving for a_x requires

$$\begin{bmatrix} d_1 \\ d_2 \\ d_3 \end{bmatrix} = [\mathbf{T}] \begin{bmatrix} a_x \\ 0 \\ 0 \end{bmatrix}$$

$$= \begin{bmatrix} \dfrac{\cos\alpha}{a_2} a_x \\ \dfrac{\sin\alpha\cos\beta}{a_1} a_x \\ \dfrac{-\sin\alpha\sin\beta}{a_3} a_x \end{bmatrix}$$

$$= \begin{bmatrix} \dfrac{\cos\alpha}{a_2} \\ \dfrac{\sin\alpha\cos\beta}{a_1} \\ \dfrac{-\sin\alpha\sin\beta}{a_3} \end{bmatrix} a_x \quad ,$$

The two sets of coordinates lie on the 'skin' of the 3D ellipsoid, and as such have a modulus equal to 1.0; this makes it convenient to solve for a_x. To determine the specific value of any one component of the coordinate, calculate the modulus of both sides and equate them. Equation (5.6) gives the modulus of the left hand side (LHS), thus we can equate this to the modulus of the right hand side (RHS) coordinates and solve for a_x:

$$a_x = 1 / \sqrt{\left(\frac{\cos\alpha}{a_2}\right)^2 + \left(\frac{\sin\alpha\cos\beta}{a_1}\right)^2 + \left(\frac{-\sin\alpha\sin\beta}{a_3}\right)^2} \tag{5.8}$$

Similarly:

$$a_y = 1 / \sqrt{\left(\frac{-\sin\alpha}{a_2}\right)^2 + \left(\frac{\cos\alpha\cos\beta}{a_1}\right)^2 + \left(\frac{-\cos\alpha\sin\beta}{a_3}\right)^2} \tag{5.9}$$

and,

$$a_z = 1/\sqrt{\left(\frac{\sin\beta}{a_1}\right)^2 + \left(\frac{\cos\beta}{a_3}\right)^2} \tag{5.10}$$

Substituting in the values for the angle parameters; $\alpha = 30°$, $\beta = 20°$, $a_1 = 150$, $a_2 = 90$, and $a_3 = 30$ yields the following apparent ranges:

$$a_x = 86.10$$
$$a_y = 79.61$$
$$a_z = 31.84$$

Remarks

Considering 3D anisotropy is not always necessary. Many geostatistical problems are 2D or are flattened by some form of geologic or stratigraphic coordinates. The concepts behind 3D geometric anisotropy are not that complex; however, the calculations are not necessarily trivial. Part of the potential complexity has to do with a variety of angle and range conventions. Practicing geostatisticians become used to dealing with different conventions and checking their specification carefully. The student of geostatistics should work through the details at least once to ensure a good understanding of anisotropy as handled in geostatistical calculations.

5.2 VARIOGRAM CALCULATION

Learning Objectives

This problem demonstrates the process of calculating experimental variograms. Getting stable, or 'well-behaved' variogram plots is a very important preliminary step to good geostatistical modeling. The configuration of real data and the sensitivity of experimental variograms to the parameters used in their computation make this challenging for early practitioners. For example, data are rarely oriented on perfect grid, so setting and fine tuning tolerance parameters to bin data is most often necessary, and not necessarily straightforward.

Background and Assumptions

The variogram is a function of distance and direction. It is approximated from data by the following calculation (Journel and Huijbregts, 1978, p.12):

$$\hat{\gamma}(\mathbf{h}) = \frac{1}{2N(\mathbf{h})} \sum_{N(\mathbf{h})} [z(\mathbf{u}) - z(\mathbf{u}+\mathbf{h})]^2 \qquad (5.11)$$

where $N(\mathbf{h})$ is the number of data pairs found for the specified lag vector, $\mathbf{h}$, $z(\mathbf{u})$ and $z(\mathbf{u}+\mathbf{h})$ are the attribute values at a location, $\mathbf{u}$, and a location separated from $\mathbf{u}$ by the lag vector $\mathbf{h}$. Of course, $2\gamma(\mathbf{h})$ is the variogram and $\gamma(\mathbf{h})$ is the semivariogram, but as mentioned previously, we follow common usage and are loose with the terminology.

Variogram inference proceeds in three main steps: (1) calculate the experimental variogram in multiple directions for a number of lags, $\mathbf{h}$, that approximately correspond to the average spacing between data, (2) interpret the experimental variogram points and supplement them with expert judgment or analogue data, and (3) fit a valid parametric model to the directional variograms in all directions. The student will have to review the many references available on variogram inference. A good overview of the topic is given in Gringarten and Deutsch (2001). The aim here is variogram calculation.

The choice of directions and parameters for experimental variogram calculation is iterative. One would start with directions chosen to be approximately aligned with the underlying directions of geological continuity if known. A simple gridded model may be constructed with inverse distance or kriging with no directional specification. Directions of continuity may become evident by identifying zones where high or low values appear connected. Sometimes, variograms are calculated in many directions over some small azimuthal increment to reveal an underlying geometric anisotropy.

Tolerance parameters are very important in variogram calculation. If the tolerance parameters are too small, then the variogram will be noisy because there are too few data pairs in each lag bin. If the tolerance parameters are too large, then the experimental points might look nearly the same in all directions because data pairs are averaged out.

The parameters for experimental variogram calculation are chosen through trial and error. Outlier data, inappropriate coordinate system, incorrect direction specification, too small a tolerance, too large a tolerance, or anything else that is poorly specified in variogram calculation will lead to poor experimental plots, with too large of an apparent nugget effect and too short a range. Attention to detail and experience leads to good interpretable variograms in most cases.

The parameters required for variogram calculation include azimuth (azm), dip, angle tolerances for these directional specifications (atol, dtol), two half-bandwidths (band1, band2), and magnitude and number of lags (Figure 5.3). Bandwidths and tolerances for angles and lag distance are search parameters that facilitate finding a reasonable number of pairs to reliably calculate the variogram in the more common case that data are irregularly spaced.

Figure 5.3 illustrates a two-step approach to calculating variograms in a 3D domain, where the first step consists of calculating a "horizontal" variogram in the plane parallel to geologic surfaces and the second step calculates a "vertical' variogram along a depth axis (e.g., parallel to drill holes).

Problem

(i) Three vertical wells spaced at 1000 m have been cored and analyzed for porosity at regular 4 m intervals. Calculate and plot the horizontal and vertical semivariograms of porosity from the dataset *5.2_part1_3welldata.dat.*

(ii) Calculate directional experimental semivariograms using the full 3D dataset *5.2_part2_allwelldata.dat.*

Note that a subset of this same data was used in Problem 3.3. Declustering is important and is necessary to establish representative univariate statistics; however, there is no systematic widely accepted approach to decluster a variogram. In general, the data pairs are equally weighted in the calculation of the variogram for a particular lag.

A variogram calculation program will be required to complete Part (ii). There are many public and commercial programs available for this purpose. The GSLIB program *gamv* is available with the supplementary materials and could be used.

Solution Plan

(i) Plot the data from *5.2_part1_3welldata.dat* on paper or on a spreadsheet. The arrangement of the data is relatively simple: three strings of data along three vertical wellbores. The area of interest is a single geological layer, such that the aspect ratio (horizontal:vertical scale) is large. The data in this example are regularly spaced to facilitate manual calculations, so angle and lag tolerances are not required, nor is it important to consider the bandwidth parameters. In practice, it will be necessary to compromise between a small tolerance for precision and a large tolerance for stability.

(ii) Now, considering a more realistic data set, determine the nominal horizontal and vertical data spacing. Consider the variations in data spacing to determine suitable lag tolerances to yield reliable experimental variograms. Determine the most likely principal directions of geometric anisotropy. As mentioned above, the data can be vizualized to determine patterns or a map could be created by inverse distance of kriging with no directional bias. Sometimes it is helpful to plot a variogram map, which is a 2D plot of the variogram values in polar coordinates for many available separation vectors, **h.** The GSLIB program *varmap* is available for this purpose. It may also be useful to plot horizontal variograms in many different directions, say at a reasonably small azimuthal increment of 15°.

Calculate the horizontal variograms. Because of the relatively wide data spacing in the horizontal direction and the density of data in the vertical direction, a narrow vertical bandwidth should be considered to get a representative horizontal variogram.

Calculate the vertical variogram. In presence of vertical wells or drill holes, the vertical variogram will be obtained by searching along individual strings of data, so bandwidths should be set small. The vertical variogram is a combination of all wells together. It would be necessary to subset the data to calculate the variogram for each well separately.

Plot the experimental variograms. In the presence of stratigraphic or tabular deposits the distance scale is very different for the horizontal and vertical directions; separate varigoram plots will be required. In other cases, directional variograms could be plotted together.

Solution

(i) The three wells are spaced 1000 m apart in the horizontal *Y*-direction. All three wells are vertical. The samples are spaced 4 m apart in the vertical direction. Given this spacing of data, the following parameters are suggested to calculate the horizontal and vertical variograms:

Suggested Parameters for Horizontal Variogram					
azimuth	0°	dip	0°	lag distance	1000 m
azimuth tolerance	0.1°	dip tolerance	0.1°	lag tolerance	0.1 m
horizontal bandwidth	0.1 m	vertical bandwidth	4.5 m	number of lags	2

Suggested Parameters for Vertical Variogram					
azimuth	0	dip	90°	lag distance	4 m
azimuth tolerance	90°	dip tolerance	0.1°	lag tolerance	0.1 m
horizontal bandwidth	0.1 m	vertical bandwidth	0.1 m	number of lags	7

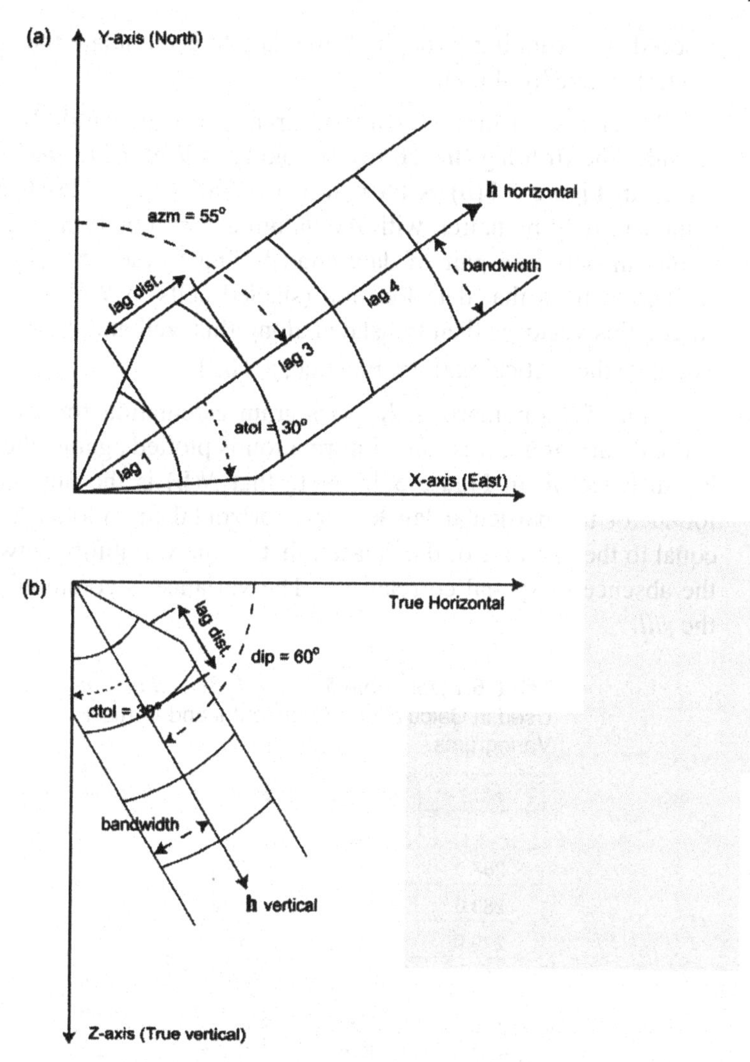

Figure 5.3 (a) Illustration of horizontal variogram calculation parameters (1) lag distance, (2) azimuth, (3) azimuth tolerance, (4) horizontal bandwidth. A vertical bandwidth for diagram (a) would be perpendicular to the plane of the page. (b) Vertical variogram calculation showing (1) lag distance (2) dip, (3) dip tolerance, and (4) bandwidth. The bandwidth parameter in (b) may be considered a "horizontal" bandwidth since it is in the plane of the variogram search. In this case, a bandwidth parameter oriented perpendicular to the search plane would entail the specification of a "vertical" bandwidth. (Redrawn from Deutsch and Journel, 1998)

Table 5.1 is an example of a spreadsheet layout of the data. For the calculation of the horizontal variogram, the shaded cells between Well 536 and 571 indicate that from location (y = 0.0, z = 271.0), there are three pairs of data contributing to the first lag (**h** =1000 m) from this data location (recall vertical bandwidth is 4.5 m, so data falling within the interval of (z−4.5m, z+4.5m) will be considered in the calculation). The calculation proceeds by scanning through all the data to accumulate the contributions to $\gamma_h(\mathbf{h}_1)$ where $\mathbf{h}_1$=4.0 m.

Calculation of the vertical variogram proceeds similarly. For example, consider the fifth lag ($\mathbf{h}_5$=20 m). Looking at Well 621, where the tail of the vector in Figure 5.3(b) is located at (y=2000.0 m, z=259.0 m), this datum value can only be paired with the datum at (y=2000.0 m, z=239.0 m). This results in only one pair of data contributing to the fifth lag of the vertical variogram from that data location (shaded cells in Well 621 in Table 5.1). Again, this vector is then translated along this well and the other two wells to calculate the vertical variogram value, $\hat{\gamma}_v(\mathbf{h}_5)$.

Table 5.2 summarizes the variogram calculation for the horizontal and vertical variogram; this same information is plotted against the corresponding lag distance, **h**, in Figure 5.4. Note that $N(\mathbf{h})$ is the number of data pairs found for the particular lag **h**. The horizontal line plotted on Figure 5.4 is equal to the variance of the dataset; that is, the variability between the data in the absence of spatial correlation. The variance is commonly referred to as the *sill*.

Table 5.1 Data from *5.2_part1_3welldata.dat* Used in Calculation of Horizontal and Vertical Variograms.

z(m)	Well 536	Well 751	Well 621
287.0		7.76	
283.0		6.02	
279.0		6.26	4.6
275.0		7.68	6.53
271.0	11.45	8.9	10.68
267.0	6.98	12.53	6.27
263.0	4.96	5.67	11.61
259.0	13.21	7.53	9.36
255.0	13.82	15.2	7.69
251.0	13.87	15.26	12.68
247.0	10.2	9.21	16.02
243.0	14.26	15.29	10.97
239.0	15.09	15.36	16.58
235.0	11.78	14.05	17.29
231.0	8.09	12.22	16.42
227.0	4.39	7.16	13.65
223.0	3.83		9.36
y(m)	0.0	1000.0	2000.0

Table 5.2 Summary of Horizontal, γ_H, and Vertical, γ_V, Semivariogram Calculations

Horizontal Variogram		
Lag No.	N(h)	γ(h)
1	78	7.31
2	38	13.55
Vertical Variogram		
Lag No.	N(h)	γ(h)
1	41	7.22
2	38	12.22
3	35	11.72
4	32	13.22
5	29	15.13
6	26	19.08
7	23	23.91
8	20	21.72

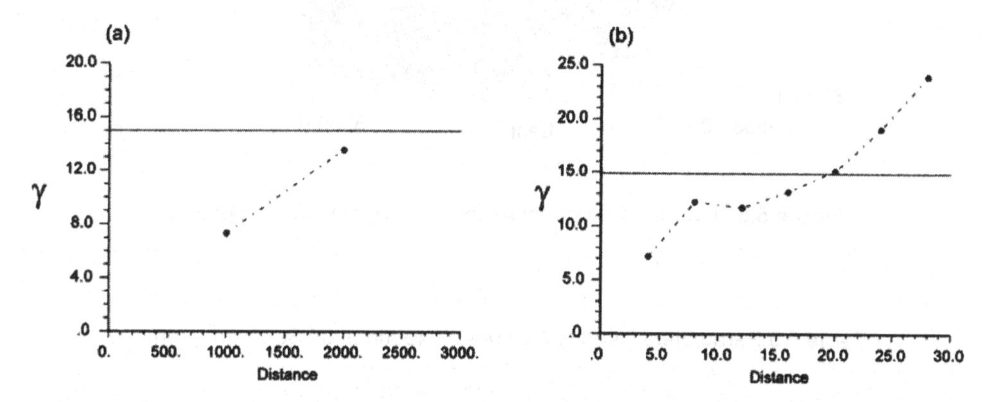

Figure 5.4 Experimental (a) horizontal and (b) vertical experimental semivariograms calculated from the data in Table 5.1. Variogram sills shown as solid horizontal line.

This first part is a quick hand calculation to understand how to proceed through the data collecting pairs for the experimental variogram. It is not realistic to attempt to calculate a horizontal variogram with only three wells. In practice, the horizontal variogram would be taken from analogue data in such a case of data sparsity.

(ii) With an adequate number of data, a variogram map will reveal any horizontal anisotropy. In this case, the maximum geologic continuity is oriented approximately NW–SE (Figure 5.5). For the subsequent step of calculating experimental semivariograms, the information contained in the horizontal variogram map guides the selection of azimuth parameters for the two horizontal directions corresponding to the directions of maximum and minimum geological continuity along stratigraphic surfaces. Specific parameters used to calculate the horizontal and vertical variograms shown here are given in Table 5.3. The experimental horizontal variograms are shown in Figure 5.6.

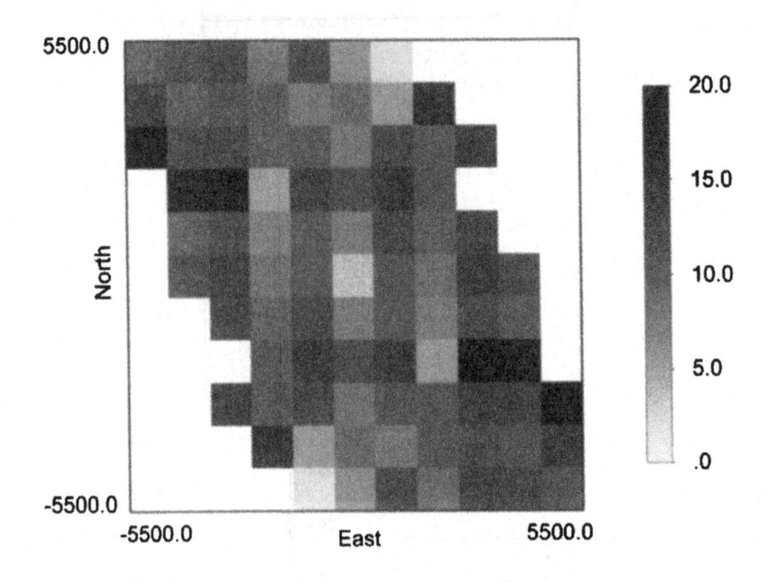

Figure 5.5 Two-dimensional horizontal variogram map of porosity.

Table 5.3 Parameters Used to Calculate Experimental Semivariograms

Parameters for Horizontal Variogram 1					
azimuth	320°	dip	0°	lag distance	550
azimuth tolerance	15°	dip tolerance	5°	lag tolerance	450
horizontal bandwidth	500 m	vertical bandwidth	5 m	number of lags	11
Parameters for Horizontal Variogram 2					
azimuth	230°	dip	0°	lag distance	550
azimuth tolerance	15°	dip tolerance	5°	lag tolerance	450
horizontal bandwidth	500 m	vertical bandwidth	5 m	number of lags	11
Parameters for Vertical Variogram					
azimuth	0°	dip	90°	lag distance	4 m
azimuth tolerance	0°	dip tolerance	10°	lag tolerance	1.5m
horizontal bandwidth	0.0 m	vertical bandwidth	10 m	number of lags	10

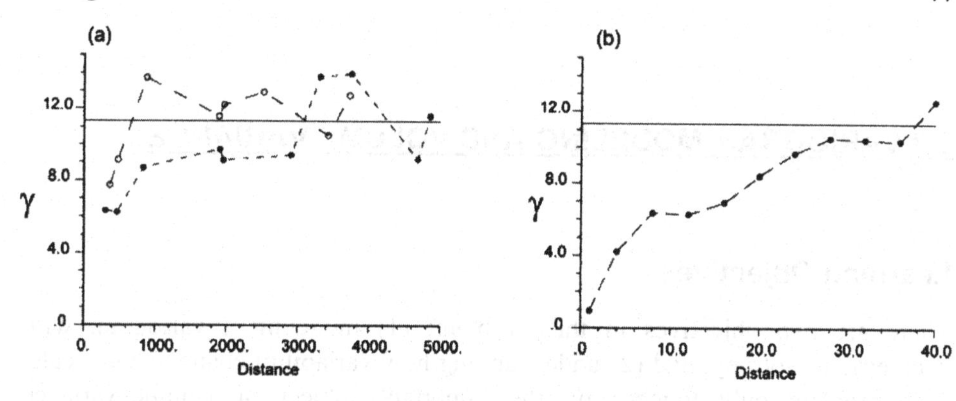

Figure 5.6 Experimental (a) horizontal and (b) vertical semivariograms of porosity obtained for the search parameters listed in Table 5.3. Variogram sills shown as solid horizontal line.

Remarks

A sensitivity study should be performed with the vertical bandwidth. A too-small value would lead to erratic unstable points. A too-large value would lead to variogram values close to the sill.

The lag separation distance should be close to the average data spacing. The number of lags used in calculating variograms is chosen such that *lag distance* times the *number of lags* equals approximately one half of the field length in a specific direction. Further details on choosing variogram calculation parameters are found in Deutsch and Journel (1998) and Gringarten and Deutsch (2000).

The inference (and subsequently modeling) of the variogram has been a longstanding area of research in geostatistics. The multitude of literature on the variogram alone is overwhelming. Some early references in this area include Davis and Borgman (1979), Cressie and Hawkins (1980), Armstrong (1984), Omre (1984), and Cressie (1985). More recently, Genton (1998a, 1998b), and Emery and Ortiz (2005) have also contributed to this literature. Given the importance of the variogram to geospatial inference, we do not doubt that the wealth and breadth of research in this area will continue.

5.3 VARIOGRAM MODELING AND VOLUME VARIANCE

Learning Objectives

There are two objectives in this problem: (1) modeling a variogram with geometric anisotropy, and (2) understanding how variability changes with scale. This problem only touches on the important subject of volume–variance relations. As the data scale increases, the variability decreases. This result may be important to understand in relation to the selection of the modeling block size given a variogram model.

Background and Assumptions

Spatial inference requires that we have a variogram–covariance value for distances and directions beyond those calculated from the available data. For this purpose, we fit a smooth function to the experimental variogram values in the three principal directions. One way to determine such a function is to perform a regression on the experimental values; however, there is no assurance that the fitted model will be positive semi-definite. For this reason, the common approach to fitting a valid model is use a linear combination of known positive semi-definite models. These models include the nugget, spherical, exponential and Gaussian models. A more extensive discussion on available models can be found in Chilès and Delfiner (1999, pp.80–104).

The data available for geologic modeling is measured on a relatively small scale given the size of the geological site being investigated. Most resource management decisions are based on larger scale trends and information. Given time and computational constraints, we create numerical models at some intermediate scale. Reconciliation of these various scales requires concepts of dispersion variance and additivity of variances. Details on these concepts are available in Journel and Huijbregts (1978), Parker (1980), Isaaks and Srivastava (1989), Glacken (1996), Goovaerts (1997), Chilès and Delfiner (1999), Journel (1999), and many other sources. A brief recall is presented below.

The average variogram $\overline{\gamma(v,V)}$ (sometimes referred to as a "gammabar" value) represents the average value of $\gamma(\mathbf{h})$ when one extremity of the vector $\mathbf{h}$ describes the volume v and the other extremity independently describes the second volume, V. The average variogram would be larger if the variogram has a nugget effect or short range, the volumes are far apart, or if the volumes are large. Often, the average variogram is calculated for $V=v$. $\overline{\gamma(v,v)}$ represents the average variability of data-scale values within the volume, v.

The dispersion variance, $D^2(V,v)$ or $\sigma^2(V,v)$, is the variance for values of volume, v, within a larger volume, V. In the classical definition of the variance, that is, the squared difference of values from their mean:

$$D^2(v,V) = E\left\{\left(\underset{\text{Support } v}{z_i} - \underset{\text{Support } V}{m_i}\right)^2\right\}$$
$$D^2(v,V) = \overline{\gamma(V,V)} - \overline{\gamma(v,v)} \tag{5.12}$$

The dispersion variance is a variance where the data are defined at one arbitrary scale (v) and the mean is defined at an arbitrarily larger scale (V). The dispersion variance is linked to the average variogram values. The following relations are important in geostatistics:

$$D^2(v,V) = \overline{\gamma(V,V)} - \overline{\gamma(v,v)}$$
$$\text{and} \tag{5.13}$$
$$D^2(\bullet,v) = \overline{\gamma(v,v)} - \overline{\gamma(\bullet,\bullet)}$$

where $D^2(\bullet,v)$ is the variance of the point scale data within the sample support volume, v. By definition, $\overline{\gamma(\bullet,\bullet)} = 0$ (Journel and Huijbregts, 1978), so $D^2(\bullet,v)$ is equivalent to the average variogram, $\overline{\gamma(v,v)}$, or gammabar, which is a measure of the variability *within* blocks of a given size, v. The gammabar value represents the mean value of the variogram $\gamma(\mathbf{h})$ as one extremity of the lag separation vector, $\mathbf{h}$, samples the domain, v, at a given location and the other extremity independently describes the same domain, v.

Problem

(i) Model the experimental semivariograms from Part (ii) of the previous Problem 5.2 using a maximum of two nested structures. All directions must be modeled using the same structures and variance contributions for each structure, but each structure may have different range parameters. The programs *vmodel* and *vargplt* are available for generating the points of a specified variogram model and plotting these against the experimental points obtained from *gamv*, which was supplied for part (ii) of the previous Problem 5.2.

(ii) Using classical volume variance relations, determine a reasonable block size for geostatistical modeling of this field for a fixed number of 1000 blocks.

As before, and throughout the problems in this book, there are many software alternatives. The GSLIB programs are provided for convenience.

Solution Plan

(i) The experimental variogram values calculated in Problem 5.2 must be fit to obtain a variogram model.

(ii) Calculate the average variogram *gammabar* for various block sizes and plot the complementary additive variance accounting for the variability between the blocks. Maximizing variability between the blocks is one reasonable criterion for the block size since an important objective in geological modeling is to represent as much heterogeneity as possible in the resulting numerical models. The program *gammabar* calculates average variogram values.

Solution

(i) The following model [Equation (5.14); Figure 5.7] is a reasonable fit to the experimental points computed from Part (ii) of the previous Problem 5.2, (Figure 5.6):

$$\gamma(\mathbf{h}) = 0.5 + 8.0Exp_{\substack{\text{ahmax=1000}\\ \text{ahmin=800}\\ \text{avert=35}}}(\mathbf{h}) + 2.8Sph_{\substack{\text{ahmax=9000}\\ \text{ahmin=1500}\\ \text{avert=35}}}(\mathbf{h}) \tag{5.14}$$

where ah_{max} is the major principal continuity direction aligned with 320° azimuth and 0° dip, ah_{min} is the minor principal continuity direction aligned in 230° azimuth and 0° dip, and a_{vert} is in the vertical direction specified by a 90° dip. Different modelers will obtain slightly different variogram models that fit the experimental data equally well. A sensitivity study could be perfromed to assess the robustness of the variogram model. The results should not be sensitive to minor changes in the variogram. If the results are not robust, then this information must be captured for decision making.

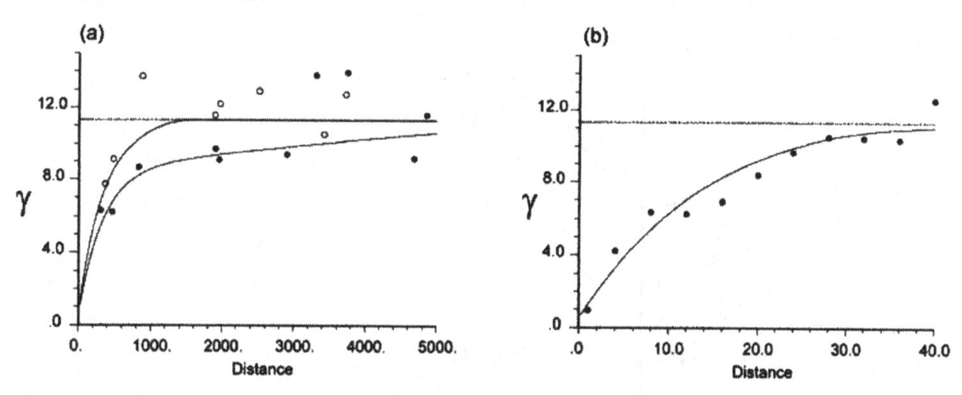

Figure 5.7 Variogram modeling results for (a) the two horizontal directions and (b) the vertical direction. The model parameters are given in Equation (5.14).

(ii) An optimal block size is one that permits maximum heterogeneity between blocks given the constraint that we will model 1000 blocks. From Equations (5.12) and (5.13), we can write

$$D^2(\bullet, A) = D^2(\bullet, v) + D^2(v, A) \tag{5.15}$$

where $D^2(\bullet, A)$ is the variability of points within the entire domain of interest, A, which is comprised of the sum of the variance for point-scale samples within the block size, v, and the variance of block of size v within A. $D^2(\bullet, A)$ is a constant equal to the variance of the sample data, which are often taken from the available point scale data. From Equation (5.12), $D^2(\bullet, v)$ is equal to the average variogram, $\overline{\gamma(v,v)}$, within blocks of support, v. Calculating $\overline{\gamma(v,v)}$ for various block sizes under the constraint that the number of modeling blocks is 1000, we seek to maximize the variability, $D^2(v, A)$, that is the modeled geological heterogeneity between blocks of size v within the area A. This result is equivalent to minimizing $\overline{\gamma(v,v)}$. Results of this calculation are summarized in Table 5.4 and Figure 5.8; once again, the results depend on the variogram and the block size.

Table 5.4 Summary of Gammabar Calculations for Different Block Sizes.

x,y size (m)	z size (m)	aspect (Lx:Lz)	nx, ny	nz	ncells	$\overline{\gamma(v,v)}=D^2(\cdot,v)$
454.55	40.00	11.36	22.00	2.00	968	7.15
555.56	26.67	20.83	18.00	3.00	972	6.51
909.09	10.00	90.91	11.00	8.00	968	6.98
1000.00	8.00	125.00	10.00	10.00	1000	7.26
1666.67	2.96	562.50	6.00	27.00	972	8.83

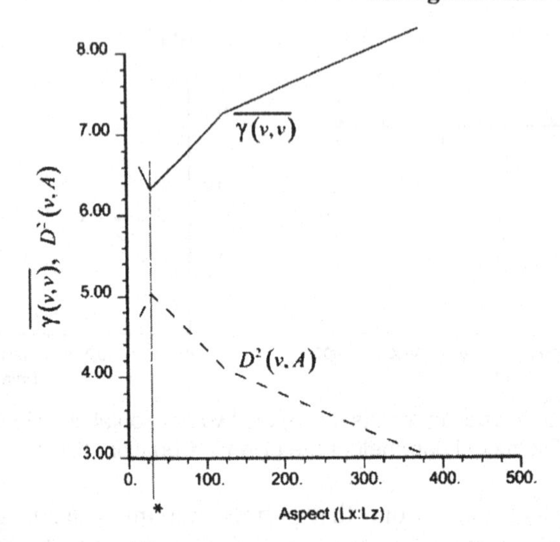

Figure 5.8 Relationship between model block dimensions and $\overline{\gamma(v,v)}$ (solid curve) and $D^2(v,A)$ (dashed curve). Optimal block size template (*) occurs at a maximum $D^2(v,A)$ and a minimum $\overline{\gamma(v,v)}$.

From Figure 5.8, $D^2(\bullet,v)$ is minimized at a block size corresponding to an aspect ratio of horizontal/vertical block size near 20; thus the variability between model blocks, $D^2(v,A)$, is maximized at this same block size aspect ratio. A reasonable horizontal block size would therefore be ~550 m and vertical block size ~25 m for a block model constrained to 1000 blocks.

Remarks

The problem of variogram modeling is often addressed by interactive software that permits dragging the variogram curve to fit the points or regression-type automatic fitting. The historical approach is based on trial-and-error where the variogram model is modified until it matches both the practitioner's experience and the available experimental points. In all cases, care must be taken to ensure that the variogram model fairly represents the anticipated variability of the geological site.

There is no unique optimal block size. Ideally, very small blocks would be used to represent the geological variability at all scales; however, this is often impractical and unnecessary. The aspect ratio of the horizontal/vertical size of the model blocks may be an important consideration. There are additional problem-specific considerations that must be considered for the block size. Most importantly, the purpose of the model must be considered.

This problem addressed the change of variance with scale; however, the entire shape of the histogram changes with scale. The variogram also changes with scale. The interested reader may refer to Oz et al. (2002), Kupfersberger et al. (1998), and Vargas-Guzman et al. (1999) for further information.

CHAPTER 6

Kriging

A defining problem of geostatistics is estimation at unsampled locations. Nearby data are considered to establish the estimates. The decision of stationarity precedes estimation; we must choose the data that are relevant to the unsampled location and we must establish the statistical parameters that describe the set of relevant data. A second decision precedes estimation; a choice of the form of the estimate. A well-established paradigm is to consider weighted linear estimates. In the early days of computer mapping, the data were assigned weights inversely proportional to their distance from the unsampled location. Such inverse distance schemes do not account for clustering of data, they do not account for the specific continuity of the variable of interest as measured by the variogram, and they do not have a clear measure of optimality. The technique of kriging was devised by Matheron and named in honor of Danie Krige following his pioneering work in the Witwatersrand gold deposits (Krige, 1951; Sichel, 1952).

Kriging is a family of linear regression tools used to calculate estimates at unsampled locations using deemed-relevant surrounding data. Prerequisites include a decision of stationarity and a variogram model. The optimality criterion for the estimate is to minimize the squared difference between the estimate and the unknown truth subject to the constraint of unbiasedness, that is, the expected value of the estimate must equal the expected value of the truth. Although the framework of optimal linear estimation was arguably developed in other disciplines (Cressie, 1990; Draper and Smith, 1966; Gandin, 1963; Goldberger, 1962; Matern, 1960), the technique of kriging within geostatistics has had a tremendous practical impact.

There are many variants of kriging based mainly on (1) different constraints to ensure unbiasedness under different models, and (2) different pre- and post

Solved Problems in Geostatistics. By O. Leuangthong, K.D. Khan, and C.V. Deutsch

transformation of the data. The names given to the common variants of kriging are unfortunate. There is nothing simple about simple kriging, there is nothing ordinary about ordinary kriging, and there is nothing universal about universal kriging. Simple, ordinary, and universal kriging will be addressed in the problems below.

The first problem establishes the theoretical and practical details of simple kriging with stationary known mean, variance, and variogram. The second problem considers the case where the mean is nonstationary under an ordinary and universal kriging framework. The last problem shows how kriging weights change depending on distance from the unsampled location and location relative to surrounding data. In particular, the notion of screening will be understood, that is, data behind other data may receive low or zero weight depending on the variogram.

6.1 STATIONARY KRIGING

Learning Objectives

The mathematics behind Kriging are central to geostatistics. The objective of this problem is to review the derivation of the kriging equations and establish the dependency on a decision of stationarity and statistical parameters, such as the variogram.

Background and Assumptions

Kriging is a method of weighted linear estimation that can be compared to traditional weighted average estimators that include inverse distance methods and moving averages. These methods are all weighted linear estimators, that is, nearby deemed relevant data are given weight according to some weighting scheme. The fundamental difference between the traditional estimators and kriging is that kriging has a well-defined minimum error variance measure of optimality (Dubrule, 1983). Consider an unsampled location $\mathbf{u}$ and n nearby sampled data locations $\mathbf{u}_\alpha$, $\alpha=1,...n$. A weighted linear estimate at the unsampled location is written as:

$$Z^*(\mathbf{u}) = \sum_{\alpha=1}^{n} \lambda_\alpha Z(\mathbf{u}_\alpha) \tag{6.1}$$

where λ_α, $\alpha = 1,...,n$ are the weights assigned to the n data. This linear combination of weights and data values is a commonly used estimator. The challenge is to assign the weights. An inverse distance weighting (IDW) scheme is perhaps the simplest, see Problem 2.3.

In the context of kriging, consider the regionalized variables at the unsampled location and all data locations as part of a random function. Under a decision of stationarity, the mean and variance at all locations are constant and given by m and σ^2, respectively. The spatial relationship between the random variable at two different locations is specified by the covariance or variogram model.

Let us proceed to establish the optimal weights based on a minimum error variance criterion. In geostatistics, this approach is referred to as kriging and there are numerous rigorous developments of kriging in the literature (Journel and Huijbregts, 1978, Isaaks and Srivastava, 1989, Goovaerts, 1997, Chilès and Delfiner, 1999). Consider the case of a stationary random field, where the mean and variance are considered to be constant and known. If we define e to be the error between the truth and an estimate, we obtain $e = Z^*(\mathbf{u}) - Z(\mathbf{u})$. Our objective is to obtain an estimate, $Z^*(\mathbf{u})$ with minimal error. A common criterion in statistics is to minimize the squared error: $e^2 = \left[Z^*(\mathbf{u}) - Z(\mathbf{u})\right]^2$. Although the true value is unknown, we are still able to minimize this squared error in expected value, that is, we minimize the error variance:

$$\sigma_E^2 = E\left\{\left[Z^*(\mathbf{u}) - Z(\mathbf{u})\right]^2\right\} \tag{6.2}$$

In a stationary field where the mean is known and is the same at every location in the field, we can simply work with the residuals, $Y(\mathbf{u}) = Z(\mathbf{u}) - m(\mathbf{u})$. The estimate then becomes

$$Y^*(\mathbf{u}) = \sum_{\alpha=1}^{n} \lambda_\alpha Y(\mathbf{u}_\alpha) \tag{6.3}$$

The Y data are known if we know the stationary constant mean $m(\mathbf{u})=m$. The error variance may be written as:

$$\sigma_E^2 = E\left\{\left[Y^*(\mathbf{u}) - Y(\mathbf{u})\right]^2\right\} \tag{6.4}$$

By expanding the squared difference term in Equation (6.4) and substituting in Equation (6.3) results in

$$\sigma_E^2 = \underbrace{C(\mathbf{0})}_{(a)} + \underbrace{\sum_{\alpha=1}^{n}\sum_{\beta=1}^{n}\lambda_\alpha\lambda_\beta C(\mathbf{u}_\alpha - \mathbf{u}_\beta)}_{(b)} - \underbrace{2\sum_{\alpha=1}^{n}\lambda_\alpha C(\mathbf{u}_\alpha - \mathbf{u}_0)}_{(c)} \tag{6.5}$$

The least-squares estimation variance of Equation (6.5) consists of three terms: (a) the variance of the data $C(\mathbf{0})=\sigma^2$; (b) the redundancy of information due to the data configuration; and (c) the distance between the data and the location being estimated. The error variance σ_E^2 is increased by the variability in the data and redundant information, but reduced by use of data that are correlated to the location being estimated. As explained in Chapter 5 (see Equation 5.1 in particular), the covariance is calculated directly from the variance and the variogram under a decision of stationarity.

The challenge is to calculate the optimal weights, that is, the values of λ_α, $\alpha=1,\ldots,n$ that minimize this error variance. Classical least-squares optimization calls for setting the partial derivatives of the estimation variance with respect to the weights, $\lambda_\alpha, \alpha=1,\ldots,n$, equal to 0. This finding leads to the following system of simple kriging equations, also referred to as the normal equations:

$$\sum_{\beta=1}^{n}\lambda_\beta C(\mathbf{u}_\beta - \mathbf{u}_\alpha) = C(\mathbf{u}_0 - \mathbf{u}_\alpha) \quad \forall \alpha = 1,\ldots,n \tag{6.6}$$

where the left-hand side (LHS) of Equation (6.6) contains all the information between the data locations, including (a) the variance, $C(\mathbf{u}_\beta - \mathbf{u}_\alpha) = C(\mathbf{0})$ when $\alpha=\beta$, and (b) the redundancy between the data values. The right-hand side (RHS) accounts for the closeness of the data to the unsampled location [component (c) in Equation (6.5)]. Solving this system of equations yields a unique set of weights that can be applied to the data values $Y(\mathbf{u}_\alpha)$, $\alpha=1, \ldots, n$ to determine the estimate $Y^*(\mathbf{u})$ via Equation (6.3). We can then add the mean to the estimated $Y^*(\mathbf{u})$ to obtain an estimate in original units $Z^*(\mathbf{u})$. This unconstrained optimization of a stationary linear estimator is called simple kriging (SK).

The solution of the normal equations to obtain the weights differentiates kriging as an optimal estimator relative to other linear estimation methods. Further, the use of the covariance function $C(\mathbf{h})$ accounts for spatial correlation structures inherent to the attribute of interest, which goes beyond a simple Euclidean distance measure. In all cases, we would like any estimate at an unsampled location to be unbiased, that is, $E\{Y(\mathbf{u}) - Y^*(\mathbf{u})\} = 0$. We can easily check that this unbiasedness property is satisfied by kriging. For the numerous properties discussed here, kriging is often referred to as the best linear unbiased estimator.

Problem

Consider the configuration of data shown in Figure 6.1. Given that three data are available at locations $\mathbf{u}_1$, $\mathbf{u}_2$, and $\mathbf{u}_3$ to estimate a value at $\mathbf{u}_0$, (i) express the estimation variance in terms of the covariance C(**h**) for this data configuration, (ii) calculate the inverse distance (ID) and inverse distance squared (ID2) weights if d=10, and (iii) compare the estimation variance obtained using classical inverse distance, inverse distance squared, and simple kriging. Assume the variable of interest is standardized and has a variogram given by

$$\gamma(\mathbf{h}) = 0.05 + 0.95\text{Sph}_{a=50}(\mathbf{h})$$

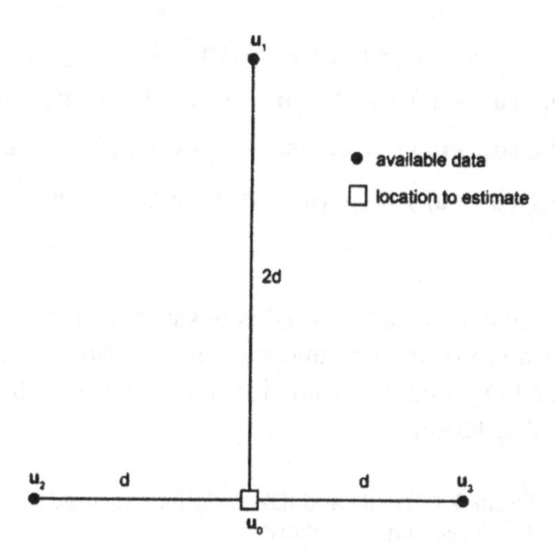

Figure 6.1 The point value $Z(\mathbf{u}_0)$ is to be estimated using three point samples $z(\mathbf{u}_\alpha), \alpha = 1,\ldots,3$.

Solution Plan

(i) Expand the estimation variance in Equation (6.5) to account for the data configuration. Although it is tedious, it may be clearer and useful for Part (ii) to write out every term in the summations.

(ii) Determine the inverse distance (ID) and inverse distance squared (ID2) weights using a power of 1.0 and 2.0, respectively (see Equation 2.17). Calculate the estimation variance for each of these sets of weights.

(iii) Determine the simple kriging weights by setting up and solving the simple kriging (or normal) equations for this configuration via Equation (6.6). Using this set of weights and the solutions to Parts (i) and (ii) of this problem, calculate the estimation variance corresponding to ID, ID2, and SK for comparison.

Solution

(i) Based on Equation (6.5), we can expand the estimation variance to account for the current data configuration of $n=3$:

$$\begin{aligned}
\sigma_E^2 &= C(\mathbf{0}) + \sum_{\alpha=1}^{3}\sum_{\beta=1}^{3} \lambda_\alpha \lambda_\beta C(\mathbf{u}_\alpha - \mathbf{u}_\beta) - 2\sum_{\alpha=1}^{3} \lambda_\alpha C(\mathbf{u}_\alpha - \mathbf{u}_0) \\
&= C(0) + \lambda_1\lambda_1 C(\mathbf{u}_1 - \mathbf{u}_1) + \lambda_1\lambda_2 C(\mathbf{u}_1 - \mathbf{u}_2) + \lambda_1\lambda_3 C(\mathbf{u}_1 - \mathbf{u}_3) \\
&\quad + \lambda_2\lambda_1 C(\mathbf{u}_2 - \mathbf{u}_1) + \lambda_2\lambda_2 C(\mathbf{u}_2 - \mathbf{u}_2) + \lambda_2\lambda_3 C(\mathbf{u}_2 - \mathbf{u}_3) \\
&\quad + \lambda_3\lambda_1 C(\mathbf{u}_3 - \mathbf{u}_1) + \lambda_3\lambda_2 C(\mathbf{u}_3 - \mathbf{u}_2) + \lambda_3\lambda_3 C(\mathbf{u}_3 - \mathbf{u}_3) \\
&\quad - 2\left[\lambda_1 C(\mathbf{u}_1 - \mathbf{u}_0) + \lambda_2 C(\mathbf{u}_2 - \mathbf{u}_0) + \lambda_3 C(\mathbf{u}_3 - \mathbf{u}_0)\right]
\end{aligned} \tag{6.7}$$

(ii) Solving for the inverse distance weights is straightforward once the distances between the location of interest and the data locations is determined. Table 6.1 summarizes the weights assigned using an inverse distance and inverse distance squared criterion.

Table 6.1 The ID and ID2 Weights Assigned to the Three Data Locations

		Weights	
Data Location	Distance to u_0	ID	ID2
1	20	0.200	0.111
2	10	0.400	0.444
3	10	0.400	0.444

(iii) Based on the SK system of equations given in Equation (6.6), we can write the following expanded system to solve for the SK weights:

$$\begin{bmatrix} C(\mathbf{u}_1-\mathbf{u}_1) & C(\mathbf{u}_1-\mathbf{u}_2) & C(\mathbf{u}_1-\mathbf{u}_3) \\ C(\mathbf{u}_2-\mathbf{u}_1) & C(\mathbf{u}_2-\mathbf{u}_2) & C(\mathbf{u}_2-\mathbf{u}_3) \\ C(\mathbf{u}_3-\mathbf{u}_1) & C(\mathbf{u}_3-\mathbf{u}_2) & C(\mathbf{u}_3-\mathbf{u}_3) \end{bmatrix} \begin{bmatrix} \lambda_1 \\ \lambda_2 \\ \lambda_3 \end{bmatrix} = \begin{bmatrix} C(\mathbf{u}_0-\mathbf{u}_1) \\ C(\mathbf{u}_0-\mathbf{u}_2) \\ C(\mathbf{u}_0-\mathbf{u}_3) \end{bmatrix}$$

The variogram is given, so the covariance is known (see Equation 5.1). Further, the covariance function is isotropic so all that is needed are the scalar distances between the data for the LHS covariance matrix, and the data to the location of interest for the RHS covariance vector. The resulting system is

$$\begin{bmatrix} 1.0 & 0.355 & 0.355 \\ 0.355 & 1.0 & 0.410 \\ 0.355 & 0.410 & 1.0 \end{bmatrix} \begin{bmatrix} \lambda_1 \\ \lambda_2 \\ \lambda_3 \end{bmatrix} = \begin{bmatrix} 0.410 \\ 0.669 \\ 0.669 \end{bmatrix}$$

Solving this system yields the weights: $\lambda_1=0.090$, $\lambda_2=0.452$, and $\lambda_3=0.452$.

To calculate the estimation variance for ID, ID2, and SK, we now consider the expanded version of the estimation variance from Part (i) given by Equation (6.7). By using the weights for ID, ID2, and SK, we obtain the following estimation variances:

$$\sigma_{ID}^2 = 0.371$$
$$\sigma_{ID2}^2 = 0.360$$
$$\sigma_{SK}^2 = 0.359$$

For this simple data configuration, there is little difference between the SK and ID2 estimators. But as expected, SK yields the lowest estimation variance of the three methods; in fact, SK will always give the minimum estimation variance (it was constructed to do just that!).

Remarks

An anisotropic correlation structure would yield different results; specifically, if the maximum correlation range is oriented in the N–S direction (in line with data $\mathbf{u}_1$), then the weight assigned to that data would be higher than the weight given to locations $\mathbf{u}_2$ and $\mathbf{u}_3$. Thus, unlike other conventional estimation schemes that only consider the geometry of the data, kriging also accounts for the spatial correlation structure of the attribute.

This feature of kriging is important and distinguishes it from other approaches. A consequence is data screening that is particular to kriging. Problem 6.3 demonstrates this phenomenon, and Problem 9.2 employs the popular Markov model for collocated cokriging that is based on this property of data screening.

6.2 NONSTATIONARY KRIGING

Learning Objectives

An implicit and fundamental assumption made in geostatistics is that the underlying statistics are invariant under translation in space. This property is called stationarity. The development of kriging in Problem 6.1 considered the case of kriging with a stationary mean, variance, variogram–covariance referred to as simple kriging (SK). Stationarity is a property of the random function construct; it is not an inherent property of actual data. In practice, real geological sites have structured and large-scale variations. Often, the mean changes regionally and in some instances the variance also changes depending on the region. The objective of this problem is to explore kriging with a nonstationary mean and introduce some common variants of kriging.

Background and Assumptions

In classical geostatistics, the decision of stationarity considers the one- and two-point cdfs and their moments to be constant over a given geological region within which a specific geostatistical model is applied (see Deutsch and Journel, 1998, p.12; Goovaerts, 1997, p.70). Natural variables often show large scale trends in the mean. After all, most geological sites chosen for study contain anomalies with areas of higher and lower values of particular interest.

In simple kriging, estimation proceeds with residuals from a specified mean: $Y(\mathbf{u})=Z(\mathbf{u})-m(\mathbf{u})$. It is the residual variable $Y(\mathbf{u})$ that is considered stationary. The mean $m(\mathbf{u})$ must be known at all locations to calculate the residual data and the estimates in original units. The mean may be set to the global mean of the declustered distribution or it could be modeled with some type of trend model. Taking the mean as a constant is straightforward. Modeling a locally varying mean is not. Kriging is used to calculate an estimate; requiring a pre- estimated mean at all unsampled locations appears circular and is practically difficult. The family of kriging techniques developed here assumes the mean is location dependent, but the estimation of the mean will be performed implicitly by the kriging.

Consider that the mean can be expressed as a linear combination of polynomials, $f_l(\mathbf{u})$, $l=0,...,L$, of the coordinates, then the mean becomes

$$m(\mathbf{u}) = \sum_{l=0}^{L} a_l f_l(\mathbf{u}) \tag{6.8}$$

If we substitute this form of the mean into the estimator of Equation (6.3), we obtain

$$Z^*(\mathbf{u}) - \sum_{l=0}^{L} a_l f_l(\mathbf{u}) = \sum_{\alpha=1}^{n} \lambda_\alpha \left[Z(\mathbf{u}_\alpha) - \sum_{l=0}^{L} a_l f_l(\mathbf{u}_\alpha) \right]$$

This form of the kriging estimate is referred to as universal kriging (UK) (Matheron, 1969; Huijbregts and Matheron, 1971). A special case of UK is ordinary kriging (OK), where $L=0$, $f_l(\mathbf{u})=1.0$ and the mean is taken as an unknown constant a_0. The form of the mean is specified by the polynomials $f_l(\mathbf{u})$, $l=0,...,L$. The precise shape and magnitude requires that the $L+1$ unknown parameters a_l, $l=0,...,L$ be locally estimated. The important contribution of UK was a scheme to implicitly calculate the a_l parameters.

The basis for the scheme is the requirement for unbiasedness of the estimate, that is, $E\{Z(\mathbf{u}) - Z^*(\mathbf{u})\} = 0$. Unbiasedness imposes linear constraints on the optimization problem. The linear constraints will be derived in the problem set. Optimization subject to linear constraints invokes the Lagrange formalism (Reklaitis et al., 1983).

Consider a function $g(x)$ that is to be maximally or minimally constrained by a function $h(x)$. Recall that for a continuous function $f(x)$ that is differentiable for all x, the maxima–minima of $f(x)$ fall on x such that $df/dx=0$. For this constrained optimization, we define a function $Q(x,\mu) = g(x) - \mu \cdot h(x)$, where μ is a Lagrange multiplier. Constrained optimization of Q optimizes $g(x)$ subject to the linear constraints of $h(x)$. We take the partial derivatives of $Q(x,\mu)$ with respect to both the x and μ values, set all derivatives equal to zero and compute the values of x that satisfy those equations.

To determine whether the calculated points are local maxima or minima, the second partial derivative of $Q(x,\mu)$ is taken with respect to x. A positive second partial derivative corresponds to a local minimum and a negative second partial derivative indicates a local maximum.

This approach is used to solve the kriging system given a nonstationary mean. This problem explores this specific scenario and requires the use of Lagrange multipliers to determine the required set of weights.

Problem

Real data are rarely stationary. Trends in the mean are a common challenge in geological and property modeling. Nonstationary kriging provides an ability to construct models that respect these trends.

(i) For nonstationary kriging, the unbiasedness property requires that $E\{Z(\mathbf{u}) - Z^*(\mathbf{u})\} = 0$. Given the form of the mean in Equation 6.8, what resulting constraints must be satisfied to ensure the unbiasedness property of kriging? For ordinary kriging, where $L=0$ and $f_l(\mathbf{u})=1$, what is the unbiasedness constraint?

(ii) Ordinary kriging is the most commonly used type of kriging in practice. It yields results that are locally accurate and honors local variations in the mean without the need to specify these local means. By using the Lagrange formalism, write the expression for the objective function that will minimize the error variance subject to the constraint developed in Part (i).

(iii) Minimize the function from Part (ii) and determine the OK system of equations used to obtain the weights.

Solution Plan

(i) The functional form of the mean is given by Equation (6.8). Express the mean of the true value $E\{Z(\mathbf{u})\}$ and the mean of the estimate $E\{Z^*(\mathbf{u})\}$ using this same form, but substitute the linear combination of data for the estimate $Z^*(\mathbf{u})$. Equate the two expressions and simplify to determine the unbiasedness constraint. Substitute $l=0$ and $f_l(\mathbf{u})=1$, to determine the OK constraint.

(ii) The Lagrange formalism requires an objective function that must have at least two components: the original function to optimize and a function for each of the constraint(s).

(iii) The OK objective function to minimize has multiple unknowns: the kriging weights and the Lagrange parameter. Thus, in order to minimize this function, we need to take the partial derivative of the objective function with respect to each of these unknowns and set it equal to zero. The resulting set of equations forms the OK system of equations.

Solution

(i) Using the same form as Equation 6.8, the mean of the RV at location $\mathbf{u}$ is written as:

$$E\{Z(\mathbf{u})\} = \sum_{l=0}^{L} a_l f_l(\mathbf{u}) \quad (6.9)$$

The estimate is a linear combination of the data. Similarly, the expression for the estimate at location **u** is written

$$\begin{aligned} E\{Z^*(\mathbf{u})\} &= E\left\{\sum_{\alpha=1}^{n} \lambda_\alpha Z(\mathbf{u}_\alpha)\right\} \\ &= \sum_{\alpha=1}^{n} \lambda_\alpha E\{Z(\mathbf{u}_\alpha)\} \\ &= \sum_{\alpha=1}^{n} \lambda_\alpha \sum_{l=0}^{L} a_l f_l(\mathbf{u}_a) \\ &= \sum_{l=0}^{L} a_l \sum_{\alpha=1}^{n} \lambda_\alpha f_l(\mathbf{u}_a) \end{aligned} \quad (6.10)$$

For unbiasedness, the constraint is that Equation (6.9) must equal Equation (6.10):

$$\begin{aligned} E\{Z(\mathbf{u})\} &= E\{Z^*(\mathbf{u})\} \\ \sum_{l=0}^{L} a_l f_l(\mathbf{u}) &= \sum_{l=0}^{L} a_l \sum_{\alpha=1}^{n} \lambda_\alpha f_l(\mathbf{u}_a) \\ f_l(\mathbf{u}) &= \sum_{\alpha=1}^{n} \lambda_\alpha f_l(\mathbf{u}_a), \quad l = 0,\dots,L \end{aligned} \quad (6.11)$$

Note that the $L+1$ equations in the last row are commonly used to ensure the equality in the middle row; however, this is not the only approach. For OK, where L=0 and $f_l(\mathbf{u})$=1, the generalized unbiasedness constraint in Equation (6.11) becomes

$$\begin{aligned} 1 &= \sum_{\alpha=1}^{n} \lambda_\alpha \\ 0 &= 1 - \sum_{\alpha=1}^{n} \lambda_\alpha \end{aligned} \quad (6.12)$$

So the sum of the weights must equal 1.0 in OK for unbiasedness. Note that with more terms, $L>0$, then we have $L+1$ constraints for Equation (6.11).

(ii) In OK, where there is only constraint Equation (6.12), the Lagrange formalism requires two components to the objective function: the estimation variance and the constraint function

$$Q=\sigma_E^2+2\mu\left(1-\sum_{\alpha=1}^{n}\lambda_\alpha\right) \tag{6.13}$$

where σ_E^2 is given by Equation (6.2). Note that the scalar 2 in front of μ is not required, but it is primarily there for convenience, when we move to optimize this function [see Part (iii)]. The reader need not include the factor 2 and should take care of the scalar coefficients when minimizing the function.

(iii) To minimize the objective function, Q, in Equation (6.13), take the partial derivative of Q with respect to the kriging weights, λ_α, and the Lagrange parameter, μ. This gives the following set of equations:

$$\begin{aligned}\frac{\partial Q}{\partial \lambda_\alpha}&=0=2\sum_{\beta=1}^{n}\lambda_\beta C(\mathbf{u}_\beta-\mathbf{u}_\alpha)-2C(\mathbf{u}_0-\mathbf{u}_\alpha)+2\mu,\quad \forall\alpha=1,\dots,n\\ \frac{\partial Q}{\partial \mu}&=0=1-\sum_{\alpha=1}^{n}\lambda_\alpha\end{aligned} \tag{6.14}$$

From Equation (6.14), the OK system of equations is obtained as:

$$\begin{aligned}\sum_{\beta=1}^{n}\lambda_\beta C(\mathbf{u}_\beta-\mathbf{u}_\alpha)-\mu&=C(\mathbf{u}_0-\mathbf{u}_\alpha),\quad \forall\alpha=1,\dots,n\\ \sum_{\alpha=1}^{n}\lambda_\alpha&=1\end{aligned} \tag{6.15}$$

Remarks

Universal kriging has great flexibility to handle different types of trends in the mean. Determining the required number of functionals $L+1$ is nontrivial; in general, the mean should be as simple as possible. Once L is determined, the optimal local estimate is the one that minimizes the error variance subject to the constraints imposed by unbiasedness. Note that the a_l parameters are not solved for directly in the common implementation discussed here. A different formalism would be required.

Also note that the variogram–covariance required in the kriging system is that between the residuals from the mean; however, we do not know the mean. Common practice is to calculate the variogram in directions and/or areas where the mean is relatively constant.

Ordinary kriging is remarkably robust for handling local trends in the mean. Journel and Rossi (1989) discussed situations when the use of OK or UK is appropriate. The OK system of equations is the SK system of Problem 6.1 with the addition of a row and column that handles the unity constraint on the weights. This result makes the extension of SK to OK and other forms of UK remarkably easy in computer implementations.

A variant of UK is Kriging with External Drift (KED) (Marechal, 1984; Chilès and Delfiner, 1999, pp. 354–360). In KED, L=1 and $f_0(\mathbf{u})$=1. The function $f_l(\mathbf{u})$ is taken from a deemed relevant variable, that is, a variable that is likely linearly related to the mean. For example, seismic travel time is likely linearly related to depth through a constant velocity. KED may be used to estimate the depth to a particular subsurface horizon using travel time as $f_l(\mathbf{u})$.

6.3 SCREENING EFFECT OF KRIGING

Learning Objectives

In general, kriging assigns closer data greater weight than farther away data. Nearby data reduce the influence of more distant data even if the distant data are within the range of correlation and candidates to receive some weight. At times, the weight to more distant data can be negative or close to zero. This commonly occurs when the data are linearly aligned with each other relative to the location of interest. The low weights applied to distant data is called data screening, and warrants the attention of geostatistical practitioners. This particular problem shows the special case of perfect screening that can occur in the case of an exponential variogram.

Background and Assumptions

The screening effect in kriging (Deutsch, 1995; Stein, 2002) refers to the phenomenon of dramatically reducing the weight assigned to a datum that is screened by another nearby datum approximately in line with the location being estimated (Figure 6.2). Note that screening occurs in configurations where two data are not only closely spaced, but approximately in line with the location being estimated. Simple clustering of data results in generally reduced weights due to the redundancy of information (see Problem 6.1), but does not necessarily constitute screening.

The exponential variogram model exhibits a perfect screen effect in the case of simple kriging with two samples collinear with the location being estimated. Perfect screening means that the screened data location receives a weight of exactly zero regardless of the spacing between the data and closeness of the two data to the unsampled location.

Problem

Demonstrate the perfect screening property of the exponential variogram model for the following data configuration:

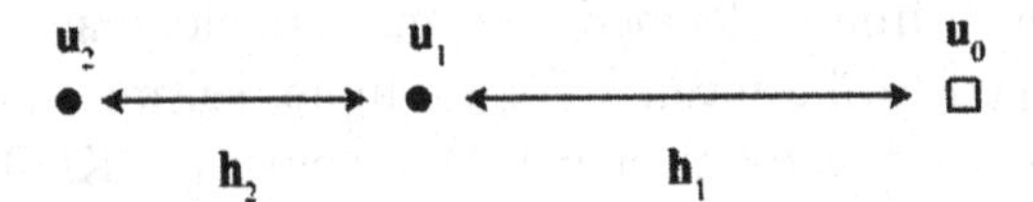

Figure 6.2 Two data, $\mathbf{u}_1$ and $\mathbf{u}_2$ fall in line with a location to be estimated, $\mathbf{u}_0$

The exponential variogram model is written

$$\gamma(\mathbf{h}) = 1 - e^{-\frac{\mathbf{h}}{a}} \tag{6.16}$$

Where a is the correlation range parameter of the model and $\mathbf{h}$ is the separation vector. Note that there is often a "3" parameter in the exponent ($-3\mathbf{h}/a$) so that the a parameter is the practical range where $\gamma(\mathbf{h})$ is 0.95 of the sill. If there is no 3, then, the a parameter is one-third of the practical range.

Solution Plan

Set up the SK equations for the two data configuration to estimate location $\mathbf{u}_0$. Reduce the system of equations to solve for the weights. It may be simplest to work this out in terms of covariances and later substitute the form of the exponential variogram. Once the exponential variogram is substituted into the equations, simplify the weight equations to show that the weight assigned to location $\mathbf{u}_2$ (Figure 6.2) is indeed zero.

Solution

The estimate at location $\mathbf{u}_0$ is given by

$$Y(\mathbf{u}) = \lambda_1 Y(\mathbf{u}_1) + \lambda_2 Y(\mathbf{u}_2) + [1 - \lambda_1 - \lambda_2]m \qquad (6.17)$$

where m is the stationary mean. The weights λ_1 and λ_2 assigned to the data at locations $\mathbf{u}_1$ and $\mathbf{u}_2$ are obtained as the solution to the simple kriging system of equations. The simple kriging equations are expressed in matrix form as:

$$\begin{bmatrix} C_{11} & C_{12} \\ C_{21} & C_{22} \end{bmatrix} \begin{bmatrix} \lambda_1 \\ \lambda_2 \end{bmatrix} = \begin{bmatrix} C_{10} \\ C_{20} \end{bmatrix} \qquad (6.18)$$

$$\begin{bmatrix} \lambda_1 \\ \lambda_2 \end{bmatrix} = [\mathbf{C}]^{-1} \begin{bmatrix} C_{10} \\ C_{20} \end{bmatrix} \qquad (6.19)$$

$$[\mathbf{C}]^{-1} = \frac{1}{C_{11}C_{22} - C_{12}C_{21}} \begin{bmatrix} C_{22} & -C_{12} \\ -C_{21} & C_{11} \end{bmatrix} \qquad (6.20)$$

$$\begin{bmatrix} \lambda_1 \\ \lambda_2 \end{bmatrix} = \frac{1}{C_{11}C_{22} - C_{12}C_{21}} \begin{bmatrix} C_{22}C_{10} - C_{12}C_{20} \\ -C_{21}C_{10} + C_{11}C_{20} \end{bmatrix} \qquad (6.21)$$

$$\lambda_1 = \frac{C_{22}C_{10} - C_{12}C_{20}}{C_{11}C_{22} - C_{12}C_{21}}$$

$$\lambda_2 = \frac{-C_{21}C_{10} + C_{11}C_{20}}{C_{11}C_{22} - C_{12}C_{21}}$$

For the data configuration considered, the elements in the RHS covariance vector and the LHS covariance matrix of Equation (6.18) are given by

$$
\begin{aligned}
C_{11} &= C_{22} = 1 \\
C_{12} &= C_{21} = e^{-\frac{\mathbf{h}_2}{a}} \\
C_{10} &= e^{-\frac{\mathbf{h}_1}{a}} \\
C_{20} &= e^{-\frac{(\mathbf{h}_1+\mathbf{h}_2)}{a}}
\end{aligned}
\tag{6.22}
$$

By substituting these values from the exponential covariance into the equations for the weights:

$$
\begin{aligned}
\lambda_1 &= \frac{e^{-\frac{\mathbf{h}_1}{a}} - e^{-\frac{\mathbf{h}_2}{a}} e^{-\frac{(\mathbf{h}_1+\mathbf{h}_2)}{a}}}{1 - e^{-\frac{\mathbf{h}_2}{a}} e^{-\frac{\mathbf{h}_2}{a}}} = \frac{e^{-\frac{\mathbf{h}_1}{a}} - e^{-\frac{(\mathbf{h}_1+2\mathbf{h}_2)}{a}}}{1 - e^{-\frac{2\mathbf{h}_2}{a}}} \\
\lambda_2 &= \frac{-e^{-\frac{\mathbf{h}_2}{a}} e^{-\frac{\mathbf{h}_1}{a}} + e^{-\frac{(\mathbf{h}_1+\mathbf{h}_2)}{a}}}{1 - e^{-\frac{\mathbf{h}_2}{a}} e^{-\frac{\mathbf{h}_2}{a}}} = \frac{-e^{-\frac{(\mathbf{h}_1+\mathbf{h}_2)}{a}} + e^{-\frac{(\mathbf{h}_1+\mathbf{h}_2)}{a}}}{1 - e^{-\frac{2\mathbf{h}_2}{a}}} = 0
\end{aligned}
\tag{6.23}
$$

Thus, we see that the weight to the more distant sample is exactly zero regardless of the precise distances involved.

Remarks

Screening is common (see Stein, 2002; Deutsch, 1994); however, perfect screening property holds only for simple kriging with an exponential variogram and for one string of collinear data. Nevertheless, close data are the most important and far data receive weights near zero. This result is practically important for techniques, such as sequential simulation, whereby a neighborhood of data are used rather than all data within the range of correlation; after the nearest 40 (or so) data, other more distant data will receive little weight even if they are correlated with the location under consideration.

The weights obtained by solving kriging are mathematically optimal subject to the problem setup. The critical choices are the data to use and the variogram model. The choice of the form for the mean is also important. Of lesser importance, but still interesting are the implicit assumptions of stationarity and ergodicity. Stationarity has been discussed. It is a required decision for the location dependence of the parameters. Ergodicity in the context of kriging is the assumption that the unsampled location and data locations are embedded in an infinite domain or, similarly, that the particular data configuration is one outcome of an infinite number with different data values.

CHAPTER 7

Gaussian Simulation

Geological heterogeneity and uncertainty are inherently linked. The unknown value at an unsampled location **u** can be described by a random variable $Z(\mathbf{u})$. The set of all random variables at the many locations of interest $\{Z(\mathbf{u}_i), i=1,...,N\}$ is referred to as a random function (RF). The number of locations is often very large: typically in the tens of millions for 3D numerical models. The N-variate multivariate probability distribution of the RF fully defines the heterogeneity and uncertainty. We denote the multivariate distribution as:

$$F(z_1, z_2, ..., z_N) = \text{Prob}\{Z_1 \leq z_1, Z_2 \leq z_2, ..., Z_N \leq z_N\}$$

Simulation amounts to drawing realizations of the N values to visualize heterogeneity and uncertainty and to transfer uncertainty through to calculated response values, such as resources and reserves. There are two challenges: (1) defining the high-dimensional multivariate distribution, and (2) drawing realizations.

An N-variate distribution cannot be defined non parametrically from data. At least 10^N data would be required to reasonably define an N-variate distribution from the data directly (see Chapter 4). A parametric model or some simplifying assumption is required. The multivariate Gaussian distribution is remarkable in its tractability; few distributions are parameterized so simply. An N-variate Gaussian distribution is fully parameterized by a vector of N mean values and a matrix of N by N covariance values. In practice, a continuous variable is transformed to follow a Gaussian histogram, the mean values are often chosen to be stationary and constant at zero, the covariance matrix is informed by a stationary covariance or variogram model. A Gaussian approach is commonly

Solved Problems in Geostatistics. By O. Leuangthong, K.D. Khan, and C.V. Deutsch

used for continuous variables because it allows tractable inference of the required multivariate distribution.

Drawing realizations from a very high dimension distribution requires some measures to make the problem tractable. One approach is turning bands (Matheron, 1973; David, 1977; Journel, 1974; Mantoglou and Wilson, 1982): 2D or 3D are reduced to a series of 1D realizations. A second approach is sequential simulation (Isaaks, 1990; Gomez-Hernandez and Journel, 1992): the multivariate distribution is decomposed by recursive application of Bayes' law and a limited neighborhood of conditioning data are retained for conditioning. There are other less common techniques. Although the different multivariate Gaussian simulation techniques are different in detail, they are fundamentally the same in the sense that they sample a multivariate Gaussian distribution. Although the decision of stationarity and the parameterization of the multivariate distribution with the required mean and covariance values are of primary importance, the details of simulating multivariate realizations are also of importance.

The first problem in this chapter establishes the link between the N-variate multivariate spatial distribution that we are trying to sample from, simple kriging and the sequential simulation approach. The second problem examines the innovative and traditional approach of conditioning unconditional simulated realizations; while this is primarily of historical interest, some simulation approaches still use this technique for conditioning. The third problem focuses on the steps required to perform sequential Gaussian simulation: a common implementation of Gaussian simulation.

7.1 BIVARIATE GAUSSIAN DISTRIBUTION

Learning Objectives

An understanding of the relationship between marginal, joint, and conditional Gaussian distributions is essential for geostatistical practitioners. Kriging for deriving conditional probability densities under a multivariate Gaussian (or multi-Gaussian) assumption is an implementation of Bayes' law. This problem demonstrates that simple kriging is equivalent to the normal equations that calculate the conditional mean and variance in the context of a multivariate Gaussian distribution.

Background and Assumptions

Bayes' law relates marginal and conditional probabilities in a multivariate context. A central problem in geostatistics is the derivation of a conditional distribution of uncertainty at an unsampled location conditional to some data. In terms of continuous probability density functions (pdfs), Bayes' law is expressed as

$$f(y|x) = \frac{f(x,y)}{f(x)} = \frac{f(x|y)f(y)}{f(x)} \quad \text{or} \quad f(x|y) = \frac{f(x,y)}{f(y)} = \frac{f(y|x)f(x)}{f(y)} \tag{7.1}$$

where $f(x,y)$ is the bivariate joint density function and $f(x|y)$ is the conditional pdf of X given the event $Y = y$. The conditional pdf $f(x|y)$ is the result of standardizing the joint pdf by the marginal distribution that governs the outcome of the conditioning event $Y = y$. Note that to be more complete and unambiguous in notation, we should denote $f(x,y)$ as $f_{x,y}(x,y)$ and $f(y)$ as $f_y(y)$, and so on; however, the simpler notation is used to reduce clutter. The ambiguity between the different functions is removed by their arguments.

The conditioning data is often of high dimension; we are interested in the conditional distribution at an unsampled location given some number of surrounding data $n>10$. It is interesting, however, to consider Bayes' law in a bivariate context first. In a spatial context, the bivariate distribution would involve one conditioning data and one unsampled location. Consider the following sketch with two random variables X and Y. The value of the X data random variable is known $X=x$. The Y random variable represents an unsampled location. The two locations are assumed to belong to the same stationary domain A. Under a decision of first-order stationarity and a prior transformation of the data to a standard normal distribution, the marginal distribution of each variable is standard Gaussian. The conditional distribution at the unsampled location $f(y|x)$ is not, however, standard Gaussian because of the conditioning by the data event $X=x$. The bivariate Gaussian distribution is parameterized by the correlation between X and Y. This correlation would be taken from the stationary variogram of the normal scores variable.

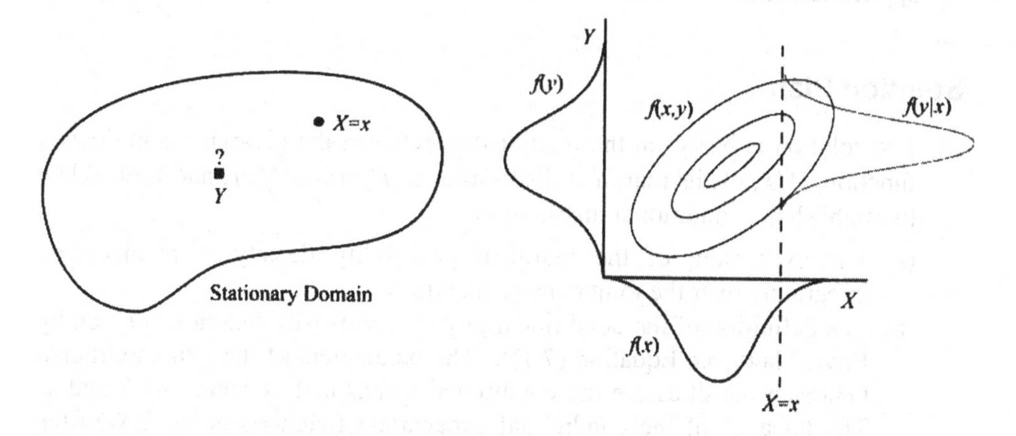

A characteristic feature of the N-variate Gaussian distribution is that all marginal and conditional distributions are Gaussian fully parameterized by mean and covariance values. The relationship between conditional moments and conditioning data is of primary concern in geostatistical calculations.

Problem

(i) Show that simple kriging under a multivariate Gaussian assumption is equivalent to Bayes' theorem. Consider the standardized (mean, $\mu = 0$; standard deviation, $\sigma = 1$) bivariate Gaussian density function as the starting point:

$$f(x,y) = \frac{1}{2\pi\sqrt{1-\rho^2}} \exp\left[-\frac{1}{2(1-\rho^2)}(x^2 - 2\rho xy + y^2)\right] \tag{7.2}$$

where

$$\rho \equiv corr(X,Y) = \frac{\sigma_{XY}}{\sigma_X \sigma_Y} \tag{7.3}$$

Calculate the mean and variance of a conditional distribution (for either $Y|X$ or $X|Y$).

(ii) Show that the sequential approach to simulation amounts to construction of a multivariate joint distribution. For this task, consider the multivariate joint distribution and the decomposition of this distribution via a recursive application of Bayes' law.

Solution Plan

(i) The solution is based on the relationship between the bivariate joint density function, $f(x,y)$, its marginal distributions, $f(y)$ and $f(x)$, and Bayes' law to establish the conditional distribution.

(a) The expression of the marginal probability density is obtained by integrating over the joint density function.

(b) The definition of the conditional probability density function is given by Bayes' law [see Equation (7.1)]. The parameters of the two conditional Gaussian densities are the conditional means and variances of X and Y. The linearity of the conditional expectation functions is the key to the solution.

(c) To better understand the proof, make a comparison between the least-squares regression line and the conditional expectation function through a cloud of data pairs from two correlated, normally distributed variables. Simple kriging is seen as a multivariate linear regression algorithm.

(ii) Consider Bayes' law in Equation (7.1) to rewrite this expression such that the joint event is expressed as a function of the conditional and marginal events. Given this same framework, write the *n*-joint distribution $f_{Z_1,\ldots,Z_n}(z_1,\ldots,z_n)$ as a function of the (*n*−*1*)-joint conditional and marginal distribution. Continue to apply Bayes' law until the joint distribution reduces to a function of *n* univariate conditional distributions.

Solution

(i) Start with the marginal distribution of a Gaussian variable, *X*,

$$f(x)=\int_{-\infty}^{+\infty} f(x,y)\,dy$$

$$=\frac{1}{2\pi\sqrt{1-\rho^2}}\int_{-\infty}^{+\infty}\exp\left[\frac{x^2-\left(y^2-2\rho xy\right)}{2\left(1-\rho^2\right)}\right]dy \tag{7.4}$$

Expand and substitute variable $u=\frac{(y-\rho x)}{\sqrt{1-\rho^2}}$, $du=\frac{dy}{\sqrt{1-\rho^2}}$ to get

$$f(x)=\frac{1}{2\pi}\exp\left(\frac{-x^2+\rho^2x^2}{2\left(1-\rho^2\right)}\right)\cdot\int_{-\infty}^{\infty}\exp\left(\frac{-u^2}{2}\right)du \tag{7.5}$$

Now recognize that $\frac{1}{\sqrt{2\pi}}\int_{-\infty}^{\infty}\exp\left(\frac{-u^2}{2}\right)du$ is the integration of the Gaussian density function that necessarily equals unity and further simplification yields the following marginal density function:

$$f(x)=\frac{1}{\sqrt{2\pi}}\exp\left(\frac{-x^2}{2}\right) \tag{7.6}$$

This marginal density function result is the standard normal distribution of variable X. The expression for the marginal distribution of Y is analogous to Equation (7.6).

The conditional distribution of Y given event $X=x$ is derived beginning from the definition of the conditional probability in Equation (7.1).

$$f(y|x)=\frac{f(x,y)}{f(x)}$$

$$f(y|x)=\frac{\frac{1}{2\pi\sqrt{1-\rho^2}}\exp\left[-\frac{1}{2(1-\rho^2)}(x^2-2\rho xy+y^2)\right]}{\frac{1}{\sqrt{2\pi}}\exp\left(\frac{-x^2}{2}\right)}$$

$$f(y|x)=\frac{1}{\sqrt{2\pi}\sqrt{1-\rho^2}}\exp\left[\frac{-y^2+2\rho xy-\rho^2x^2}{2\left(1-\rho^2\right)}\right] \quad (7.7)$$

$$f(y|x)=\frac{1}{\sqrt{2\pi}\sqrt{1-\rho^2}}\exp\left[-\frac{1}{2}\left(\frac{y-\rho x}{\sqrt{1-\rho^2}}\right)^2\right]$$

The conditional distributions, like the marginals of the bivariate Gaussian density, are also Gaussian. From the result of Equation (7.7), the expressions for the mean and the variance of the conditional Gaussian probability density $f(y|x)$ are, respectively,

$$\mu_{Y|X=x}=\rho x \quad (7.8)$$

$$\sigma^2_{Y|X=x}=1-\rho^2 \quad (7.9)$$

Equation (7.8) gives the expression for the conditional expectation function $E\{Y|X=x\}$ for the bivariate distribution $f_{x,y}(x,y)$ in Equation (7.2). Two characteristics of the conditional expectation and variance, respectively, are the linearity of the conditional expectation and the independence of the conditional variance on the value of the conditioning event, $X=x$.

Now consider simple kriging of a standard RV Y, given a nearby data X separated by a lag distance of **h**. Also assume that the covariance function C(**h**) is known. Since the RVs are standard, the covariance function, C(**h**) is

kriging is used for conditioning. This approach to conditioning is of theoretical and practical interest; it is still used today in turning bands simulation, as well as any matrix, spectral, and sometimes object-based modeling approaches. The objective of this problem is to (1) understand how conditioning by kriging works, and (2) show the simplicity of the method in practice.

Background and Assumptions

Sequential simulation could be interpreted in one of two ways (1) recursive sampling of conditional distributions, or (2) adding a random error to a simple kriging estimate to introduce the correct variability in an unbiased manner. The first view of sequential simulation was introduced in the previous problem. The second view arises from the common implementation. In practice, locations are visited in random order and a random error, $R(\mathbf{u})$, is added to the smoothly varying estimator, $Z^*(\mathbf{u})$, obtained by simple kriging with nearby data. This amounts to draw from the correct local conditional distribution. The correct stationary variance is then observed at each location. Simulation adds to the kriged field a realization of the random function error, $R(\mathbf{u})$, having zero mean and a variance equal to the simple kriging variance, σ^2_{SK}:

$$Z(\mathbf{u}) = Z^*(\mathbf{u}) + R(\mathbf{u}) \tag{7.15}$$

The resultant simulated field, $\{Z_s(\mathbf{u}), \mathbf{u} \in A\}$, has the correct variance and correct covariance between all spatial locations. Each realization can be interpreted as a legitimate sample from the multivariate Gaussian distribution between all simulated and data locations.

This process of *conditioning by kriging* is summarized as (Journel and Huijbregts, 1978, pp. 494–498):

$$\begin{aligned} z_{cs}(\mathbf{u}) &= z^*(\mathbf{u}) + \left[z_s(\mathbf{u}) - z_s^*(\mathbf{u}) \right] \\ r(\mathbf{u}) &= z_s(\mathbf{u}) - z_s^*(\mathbf{u}) \end{aligned} \tag{7.16}$$

where $z_{cs}(\mathbf{u})$ is a conditional simulation. Conditioning to the data is achieved by adding the simulated error field, $r(\mathbf{u})$, which is known exactly at the data locations, to the field of kriged estimates, $z^*(\mathbf{u})$. The correlated error field, $r(\mathbf{u})$, is decomposed as an unconditional simulation, $z_s(\mathbf{u})$, having the correct covariance function, $C(\mathbf{h})$, between all locations, and a SK of the unconditionally simulated values retained at the locations of the conditioning data, $z_s^*(\mathbf{u}_\alpha)$, $\alpha = 1,\ldots,n$. The resultant conditional simulation, $z_{cs}(\mathbf{u})$, honors the data at the data locations because of the exactitude of kriging.

Note further that Equation (7.16) can also be re-expressed as:

$$\begin{aligned} z_{cs}(\mathbf{u}) &= z_s(\mathbf{u}) + \left[z^*(\mathbf{u}) - z_s^*(\mathbf{u})\right] \\ &= z_s(\mathbf{u}) + \Delta z^*(\mathbf{u}) \end{aligned} \tag{7.17}$$

where $\Delta z^*(\mathbf{u}) = z^*(\mathbf{u}) - z_s^*(\mathbf{u})$, that is, the difference between the SK of the data and the SK of the subset of $z_s(\mathbf{u})$ retained at the data locations. This re-expression shows that we can obtain a simulated value by (1) performing unconditional simulation of the field to obtain $z_s(\mathbf{u})$, and (2) kriging of the difference between the unconditional simulated value retained at the data locations and the data values themselves.

Problem

(i) Show that the covariance, $C(\mathbf{h})$, of the conditional simulation, $z_{cs}(\mathbf{u})$, is correct, as we know it is for the unconditional realization, $z_s(\mathbf{u})$. For this proof, consider the covariance of the conditionally simulated values between two unsampled locations, $\mathbf{u}_1$ and $\mathbf{u}_2$.

(ii) By using the 1D dataset, *7.2_transect.dat*, generate a conditional simulation using the conditioning by kriging method and plot the results.

Solution Plan

(i) Express the conditionally simulated values at $\mathbf{u}_1$ and $\mathbf{u}_2$ in the form of Equation (7.17). Write the expression for the covariance of these two simulated values in expected value notation. For each expected value term, expand and simplify the form of the expected value by using the expanded form of the simulated value as rewritten using Equation (7.17).

(ii) For this part, two approaches are possible. The first approach consists of the following steps:

(a) Generate a kriged field of estimates by performing SK on the Gaussian-transformed dataset in *7.2_transect.dat*.

(b) Generate an unconditional Gaussian simulation, retaining the values at the data locations.
(c) Perform a second simple kriging using these values.
(d) Take the difference between the fields of (b) and (c) and add the result to the kriged field from (i).

An alternative solution method, following Equation (7.17), is
(a) Generate an unconditional Gaussian simulation and calculate the difference between the data and the simulated values at the data locations and retain these values.
(b) Perform SK using these values.
(c) Add this SK field to the unconditional simulation.

Solution

(i) Consider two unsampled locations, $\mathbf{u}_1$ and $\mathbf{u}_2$, such that

$$\begin{aligned} z_{cs}(\mathbf{u}_1) &= z_s(\mathbf{u}_1) + \Delta z^*(\mathbf{u}_1) \\ z_{cs}(\mathbf{u}_2) &= z_s(\mathbf{u}_2) + \Delta z^*(\mathbf{u}_2) \end{aligned} \tag{7.18}$$

The covariance between $z_{cs}(\mathbf{u}_1)$ and $z_{cs}(\mathbf{u}_2)$ is

$$Cov\{z_{cs}(\mathbf{u}_1), z_{cs}(\mathbf{u}_2)\} = E\{z_{cs}(\mathbf{u}_1) \cdot z_{cs}(\mathbf{u}_2)\} - m^2_{z_{cs}} \tag{7.19}$$

We can assume that the mean of the conditionally simulated values, $m_{z_{cs}}$, is equal to 0 without any loss of generality, thus the covariance becomes

$$\begin{aligned} Cov\{z_{cs}(\mathbf{u}_1), z_{cs}(\mathbf{u}_2)\} &= E\{[z_s(\mathbf{u}_1) + \Delta z^*(\mathbf{u}_1)][z_s(\mathbf{u}_2) + \Delta z^*(\mathbf{u}_2)]\} \\ &= E\{z_s(\mathbf{u}_1) z_s(\mathbf{u}_2)\} + E\{\Delta z^*(\mathbf{u}_2) z_s(\mathbf{u}_1)\} \\ &\quad + E\{\Delta z^*(\mathbf{u}_1) z_s(\mathbf{u}_2)\} + E\{\Delta z^*(\mathbf{u}_1) \Delta z^*(\mathbf{u}_2)\} \end{aligned} \tag{7.20}$$

Since $\Delta z^*(\mathbf{u})$ is simply the addition of two kriged fields, we can express it as a kriging estimator:

$$\begin{aligned}\Delta z^*(\mathbf{u}) &= z^*(\mathbf{u}) - z_s^*(\mathbf{u}) \\ &= \sum_{\alpha=1}^{n} \lambda_\alpha z(\mathbf{u}_\alpha) - \sum_{\alpha=1}^{n} \lambda_\alpha z_s(\mathbf{u}_\alpha) \\ &= \sum_{\alpha=1}^{n} \lambda_\alpha \left[z(\mathbf{u}_\alpha) - z_s(\mathbf{u}_\alpha) \right] \\ &= \sum_{\alpha=1}^{n} \lambda_\alpha \left[z^*(\mathbf{u}_\alpha) - z_s^*(\mathbf{u}_\alpha) \right]\end{aligned} \quad (7.21)$$

Substituting Equation (7.21) into the second term on the RHS of Equation (7.20):

$$\begin{aligned}&E\left\{ \sum_{\alpha=1}^{n} \lambda_\alpha \left[z^*(\mathbf{u}_\alpha) - z_s^*(\mathbf{u}_\alpha) \right] \cdot z_s(\mathbf{u}_1) \right\} \\ &= \sum_{\alpha=1}^{n} \lambda_\alpha \left[E\left\{ z^*(\mathbf{u}_\alpha) z_s(\mathbf{u}_1) \right\} - E\left\{ z_s^*(\mathbf{u}_\alpha) z_s(\mathbf{u}_1) \right\} \right]\end{aligned} \quad (7.22)$$

Since the covariance between a kriged estimate at a data location and any other location, that is, $\mathbf{C(u_\alpha,u)}$, is correct, recognize each term on the right of Equation (7.22) as these covariances, thus

$$\begin{aligned}&E\left\{ \sum_{\alpha=1}^{n} \lambda_\alpha \left[z^*(\mathbf{u}_\alpha) - z_s^*(\mathbf{u}_\alpha) \right] \cdot z_s(\mathbf{u}_1) \right\} \\ &= \sum_{\alpha=1}^{n} \lambda_\alpha \left[C(\mathbf{u}_\alpha, \mathbf{u}_1) - C(\mathbf{u}_\alpha, \mathbf{u}_1) \right] \\ &= 0\end{aligned} \quad (7.23)$$

The third term on the RHS of Equation (7.20) is expanded in a similar fashion to the same result of Equation (7.23).

Now consider the last term on the RHS of Equation (7.20). We can substitute Equation (7.21) into this term to give

$$
\begin{aligned}
& E\left\{\sum_{\alpha=1}^{n} \lambda_\alpha \left[z^*(\mathbf{u}_\alpha) - z_s^*(\mathbf{u}_\alpha)\right] \cdot \sum_{\beta=1}^{n} \lambda_\beta \left[z^*(\mathbf{u}_\beta) - z_s^*(\mathbf{u}_\beta)\right]\right\} \\
&= \sum_{\alpha=1}^{n} \sum_{\beta=1}^{n} \lambda_a \lambda_\beta \begin{bmatrix} E\{z^*(\mathbf{u}_\alpha) z^*(\mathbf{u}_\beta)\} - E\{z^*(\mathbf{u}_\alpha) z_x^*(\mathbf{u}_\beta)\} \\ -E\{z_x^*(\mathbf{u}_\alpha) z^*(\mathbf{u}_\beta)\} + E\{z_x^*(\mathbf{u}_\alpha) z_x^*(\mathbf{u}_\beta)\} \end{bmatrix} \\
&= \sum_{\alpha=1}^{n} \sum_{\beta=1}^{n} \lambda_a \lambda_\beta \left[C(\mathbf{u}_\alpha, \mathbf{u}_\beta) - C(\mathbf{u}_\alpha, \mathbf{u}_\beta) - C(\mathbf{u}_\alpha, \mathbf{u}_\beta) + C(\mathbf{u}_\alpha, \mathbf{u}_\beta)\right] \\
&= 0
\end{aligned}
\tag{7.24}
$$

Finally, by substituting the results of Equations (7.23) and (7.24) back into Equation (7.20), we get

$$
\begin{aligned}
Cov\{z_{cs}(\mathbf{u}_1), z_{cs}(\mathbf{u}_2)\} &= E\{z_s(\mathbf{u}_1) z_s(\mathbf{u}_2)\} \\
&= Cov\{z_s(\mathbf{u}_1), z_s(\mathbf{u}_2)\}
\end{aligned}
\tag{7.25}
$$

Showing that indeed, the covariance between the conditionally simulated values is equal to the covariance between unconditionally simulated values, which we know is correct.

(ii) The results of conditioning by kriging for the given data set are shown in Figure 7.1. The plots shown illustrate the solution steps outlined in the second alternative of the Solution Plan. An unconditional simulation is first performed. At data locations, the difference between the unconditionally simulated values and the original data values are then calculated. Simple kriging is then performed using these calculated differences. The resulting kriged field is then added to the unconditionally simulated field to obtain the conditional simulated values at all locations.

Remarks

Simple kriging (SK) is also known as the normal equations [see Problem 6.1 Equation 6.6]. SK provides the theoretically correct conditional mean and variance. It is the correct approach regardless of the interpretation paradigm (i.e., drawing from the conditional distribution or adding a random component). The input variogram (or covariance) will be reproduced in simulation when SK is used. Interestingly, the fact that SK leads to reproduction of the input variogram is not necessarily based on the multivariate Gaussian distribution. The so-called simple kriging principle is that sequential simulation from any shape conditional

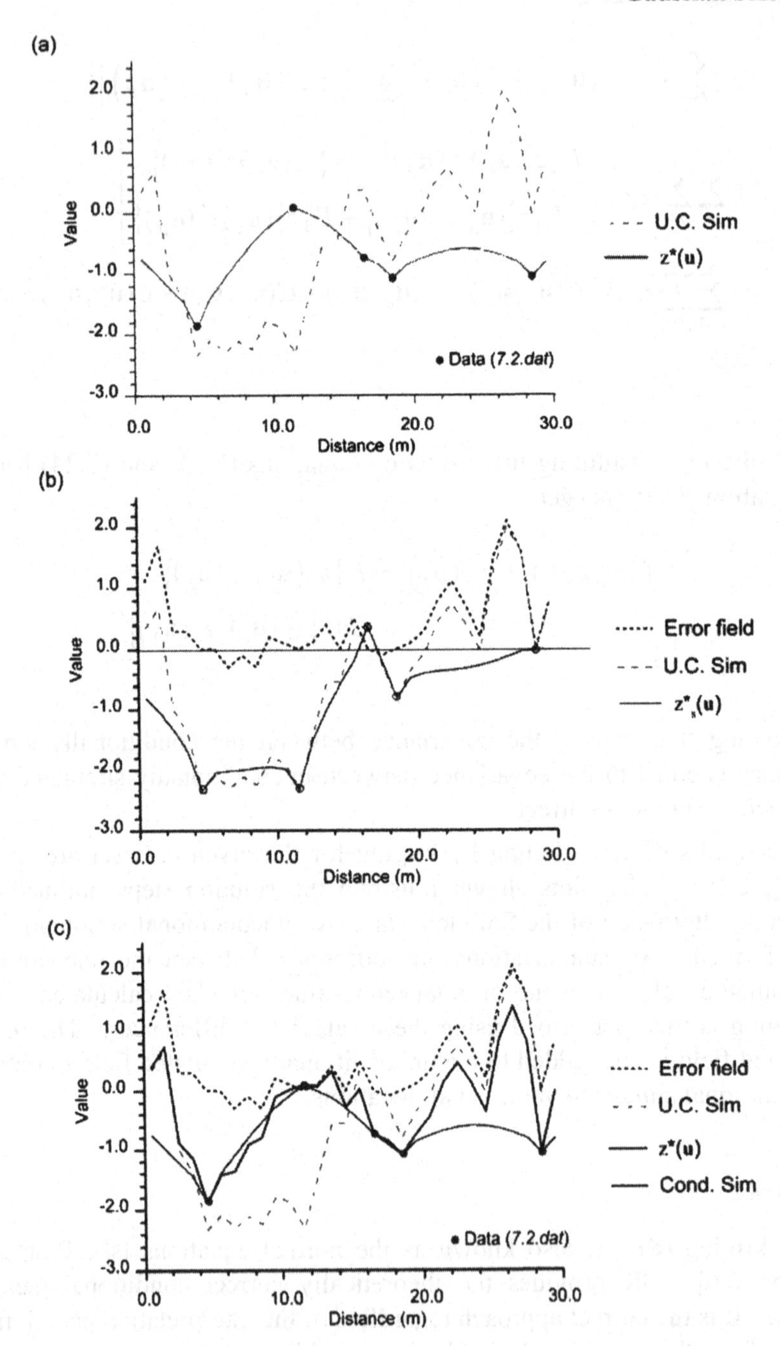

Figure 7.1 (a) Simple kriging ($z^*(\mathbf{u})$) of the data set in *7.2_transect.dat* and an unconditional sequential Gaussian simulation (U.C. Sim); (b) Simulated error field (black dashed curve) obtained by taking the difference between a Simple Kriging ($z_s^*(\mathbf{u})$) of the values retained (open circles) from the unconditional simulation at the locations of the data; (c) Final resultant conditional simulation obtained by adding the realization of the error field to the original kriging ($z^*(\mathbf{u})$) of the data.

distribution, constrained to the SK mean and variance, will lead to the correct mean, variance and variogram (covariance) values (Journel, 1994).

Although conditioning by kriging does not necessarily require a multivariate Gaussian assumption, some distribution must be assumed. Also, the Central Limit Theorem (CLT) (see Problem 4.1) tells us that the results will be Gaussian to some extent since we are adding randomly drawn values. The congeniality of working in Gaussian space is that we only need to know the conditional mean and variance (obtained via SK) in order to determine the full distribution of uncertainty about that particular location. Thus while Gaussian variables are not required *a priori*; in practice, a Gaussian transform of the data is used to enable multi-Gaussian simulation. The appropriateness of the Gaussian assumption is left to the discretion of the modeler (see Problem 8.2).

7.3 GAUSSIAN SIMULATION

Learning Objectives

Local uncertainty is defined analytically under a multivariate Gaussian distribution; SK defines the local conditional moments then quantiles of the local conditional distribution can be back transformed to provide a local (data or point scale) distribution of uncertainty. Simulation is not required to define local uncertainty. The place for simulation is when we are interested in uncertainty of more than one location simultaneously. This problem demonstrates that calculating point-wise uncertainty at all locations of a spatial domain is not equivalent to calculating the joint or multivariate uncertainty over the domain. A second objective of this problem is to appreciate that the joint distribution of kriged estimates is far from that of the random function.

Background and Assumptions

Geostatistical algorithms aim at modeling a conditional probability distribution, and hence a characterization of the uncertainty in the geological attribute(s) prevailing at each location in a spatial domain. These probability distributions are conditioned by the data set (n), sampling the attribute, Z over the domain. The desired conditional cumulative distributions (ccdfs) have the form:

$$F\left(\mathbf{u};z\,|\,(n)\right)=\text{Prob}\left\{Z(\mathbf{u})\leq z\,|\,(n)\right\} \qquad (7.26)$$

Expression (7.26) models the uncertainty about the unsampled value $z(\mathbf{u})$, but it does not provide any information about the joint uncertainty between several locations. This joint uncertainty model is obtained by simulating from the set of univariate ccdfs of the type in Equation (7.26), but with successively increasing conditioning, as shown in Problem 7.1.

Stochastic simulation from the N ccdfs of type (7.26) over the discretized domain results in a realization of the joint uncertainty model. Multiple realizations are generated in order to obtain a representative joint characterization of local uncertainty in the geological attribute(s) of interest. A given realization reproduces the bivariate statistics captured and modeled through the variogram, and so reproduces the characteristics of the geospatial heterogeneity that are resolved by two-point correlation.

When the ccdfs (7.26) are Gaussian, the multivariate probability model is fully characterized by the kriging estimate and kriging variance corresponding to the mean and variance of the ccdf at each location, $\mathbf{u}$. This simplicity leads to the widespread use of Gaussian simulation algorithms. This may also lead to the belief that a field of kriged estimates is sufficient to model spatial variability. However, the field of kriged estimates is only part of the model of spatial variability; and does not suffice as a complete characterization of spatial uncertainty.

The covariance between the kriged estimates and the neighboring data used to derive those estimates is correct; however, the covariance between the kriged estimates is incorrect. The smoothness of kriging is expressed as $Var\{Y^*(\mathbf{u})\} = C(0) - \sigma_{SK}^2(\mathbf{u})$; in other words: the variance of the kriged estimate is less than the variance of the data by an amount equal to the simple kriging variance. The way to correct for this smoothness is to add back this amount of variance by simulating from a local ccdf that is centered about the kriged estimate with a variance derived from simple kriging.

The consequence of not simulating, and considering only the field of kriged estimates is that any transfer of uncertainty based on that model of spatial variability will be potentially biased and, in general, under-represent the range of uncertainty. A simple uncertainty transfer function is a block averaging, or upscaling of gridded data. It is common practice to build a small scale spatial uncertainty model and average the values to a coarser discretization. The goal might be estimating the variance of average grade in blocks of a selected size, or reducing a model to a computationally practical number of cells. Block averaging may be via a linear average (i.e., an arithmetic average) or some nonlinear power average of the type:

$$\overline{Z}(\mathbf{u}) = \left(\frac{1}{n} \sum_{i=1}^{n} z_i^{\omega} \right)^{\frac{1}{\omega}}$$

where ω is an exponent in [−1,+1], with the geometric average obtained with ω = 0 at the limit as $\bar{Z}(\mathbf{u})=\left(\frac{1}{n}\prod_{i=1}^{n}z_i\right)$. Block averaging is linear for mass and volume fractions; however, permeability, rate constants, and other variables used in the earth sciences do not average linearly.

Problem

Use the data file *7.3_2DBHdata.dat* to generate a 2D field of kriged estimates and a simulated realization using simple kriging. Use the following variogram model parameters to generate both models:

$$\gamma(\mathbf{h}) = 0.37Sph_{\substack{ah1=4000\\ah2=1000}}(\mathbf{h}) + 0.63Sph_{\substack{ah1=4000\\ah2=3000}}(\mathbf{h}) \tag{7.27}$$

where *ah1* is the range in the 45° azimuth direction and *ah2* is the range in the 135° azimuth direction. Generate these models on a domain of size 15,000 x 15,000 m using a discretization of 50 m x 50 m blocks. Block average both the kriged and simulation model to 1500 m x 1500 m blocks.

You can use any geostatistical software for this exercise. Any simulation procedure can be used to sample the local ccdfs; including matrix approaches (LU decomposition); turning bands; or any other means of unconditional simulation followed by a kriging step for the conditioning (see Problem 7.2). The following GSLIB programs are supplied: *gam.exe* for calculating the variogram on the gridded models for part (i) of this problem; *kb2d.exe*, a 2D kriging program; and *sgsim.exe* for sequential Gaussian simulation.

The program *blockavg.exe* is included in the supplementary materials for use in the block averaging component of this problem. The option to use different types of averaging based on a power model is available. Perform the checks by setting the averaging exponent to −1, 0, and 1 corresponding to the harmonic, geometric and arithmetic averages, respectively.

Compare the original input data to both the kriged and simulated models by

(i) The histogram and variograms of the kriging and simulation model against the statistics calculated from the data,

(ii) The histograms after block-averaging the kriged and simulated fields to the point-scale data histogram,

(iii) The proportion of values falling above a threshold of 12.0 data units for the original kriged and simulated models and the block-averaged results.

Solution Plan

(i) Perform SK of the data over a domain of size 15,000 x 15,000 m using a discretization of 50 m x 50 m blocks using the supplied variogram model. Ensure that the search radius used in the kriging is commensurate with the variogram ranges; a good rule of thumb is to use a search radius that is at least the maximum range of the variogram model. Note that the variogram model given is the variogram of the normal score transformed data, so the kriging should be performed on the transformed data. Back transform the kriged estimates in order to compare the histograms, but retain the normal scores kriged estimates to compare the variograms. Note further that the back transform of kriged estimates from Gaussian units requires a numerical integration approach to obtain kriged estimates in original units (Verly, 1983; Verly, 1984). This is because the mean in Gaussian units generally does not correspond to the mean in original data units, unless the distributions are symmetric. The program *postmg* is available in the supplementary materials for this back transform.

Run a sequential Gaussian simulation program on the same data and domain to generate a single realization. Here too, run one simulation on the normal score transformed data without a back-transform after the simulation to compare the variograms, and run a second simulation on the original data including a back transform to original data space to compare the histograms. Whatever software is used, ensure that the random path between the two realizations is the same (i.e., by using the same random number seed) so that the impact of the path is removed from the comparison. Alternatively, run one simulation using the normal score transformed data, and back transform the entire realization to original units. In this case, a straight back transform using only the transformation table between the original data untis and the normal-scores values is correct. The program *backtr* is available for this direct back transform if needed.

Compute the normal scores semivariograms of the gridded kriged estimates and the simulated values and compare these experimental variograms to the input (Gaussian) variogram model, Equation (7.27).

Plot the histograms of the back transformed kriged estimates and the simulated values and compare these to the histogram of the data. Note the reduction in the variance of the kriged estimates. Also note the shape changes in the histograms, which may be easiest to see via quantile–quantile plots (see Deutsch, 2002, p.44).

(ii) Upscale the models to a cell size of 1500 x 1500 m using arithmetic, geometric, and harmonic averaging. Note (a) the general reduction in the variance of the upscaled results relative to the point scale gridded data; and

(b) the differential variance reduction between the point scale and upscaled kriged and simulated values.

(iii) Plot the cumulative histograms of the point-scale reference data and gridded models and compare the proportions of the data above the threshold value of 12.0. The differences observed here are commensurate with the changes observed in the variance and shape of the histograms in the previous step.

Solution

(i) The smoothness of the kriged estimates is apparent in the comparison of gridded results shown in Figure 7.2. This smoothness quality of kriging is reinforced when we consider the histogram comparisons in Figure 7.3 and variogram reproduction checks (Figure 7.4) between the kriged and simulated fields, respectively. In Figure 7.3(b), the smoothness of kriging is shown by the reduced variance of the kriged estimates. On the other hand, the distribution from simulation is similar to the data not only in statistics, but also in shape [Figure 7.3(c)]. The smoothness of the kriged estimates is also apparent from exaggerated variogram ranges in the principal directions of continuity, a Gaussian character near the origin or the variogram plot, and a reduced sill [Figure 7.4(a)].

(ii) The impact of joint uncertainty is apparent when we consider a transfer function. The results after applying a harmonic and arithmetic average are shown in Figure 7.5. Note that the mean and variance of the upscaled simulated results, as well as the shape of the histogram of upscaled blocks change with the different transfer functions. On the other hand, the smoothness of the input kriged estimates results in much less variability when processed the same way. This results in an understatement of uncertainty when we consider only the kriged model.

(iii) An alternative transfer function that is nonlinear is the use of a threshold. If we apply a threshold value of 12.0 to the original data, we find that 13% of the data lies above this threshold. Applying the same threshold to the kriged model at the fine scale shows that only 5% of the model falls above the threshold. Similarly, applying this cutoff value to the simulated model shows that 15% of the realization falls above the cutoff. Clearly, the kriged model results in a significant underestimation of the desired proportion. This type of transfer function is common and important in reserve evaluation. Thus the bias resulting from kriging due to the smoothness effect is a significant issue in resource/reserve assessment.

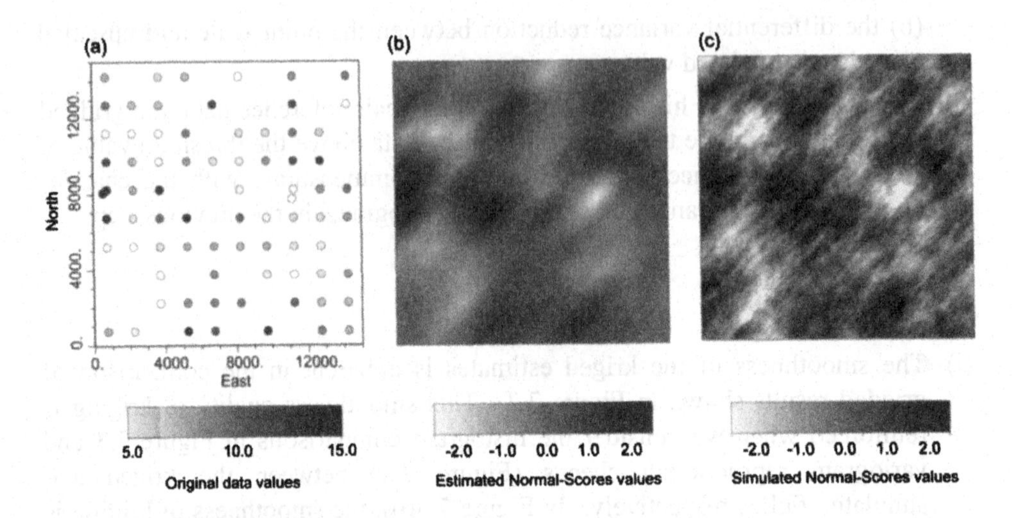

Figure 7.2 (a) Location map of the data in the file *7.3_2DBHdata.dat*, (b) gridded kriged estimates in normal-score transformed space, and (c) a simulated conditional realization also in normal-score space.

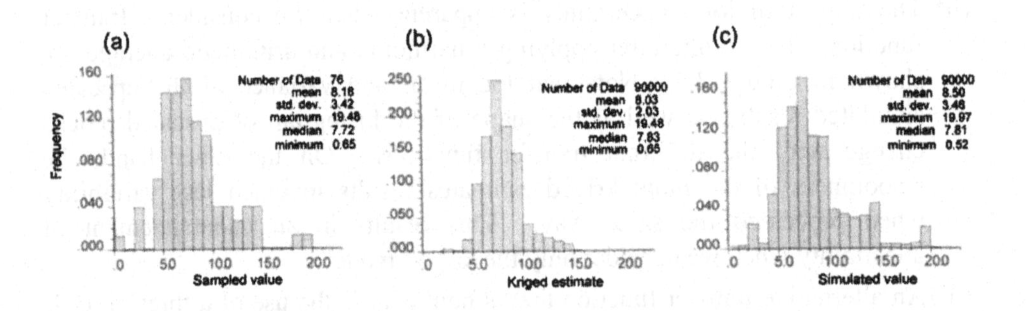

Figure 7.3 Histogram of (a) sample data from the data set *7.3_2DBHdata.dat*, (b) gridded kriged estimates, and (c) simulated grid of values. Note the significant reduction in variance reported from the kriged estimates.

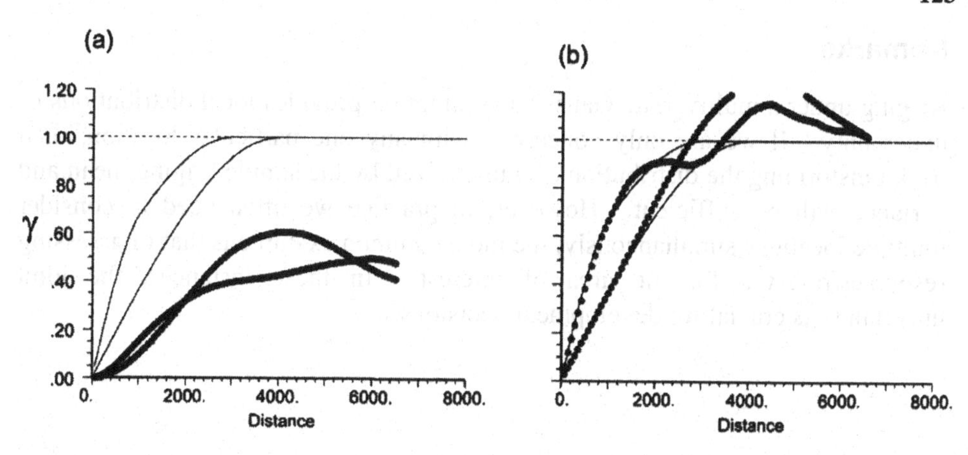

Figure 7.4 Directional semivariograms of (a) kriged estimates (points), and (b) simulated normal-scores results (points) compared to the input variogram model (solid lines) given in Equation (7.27).

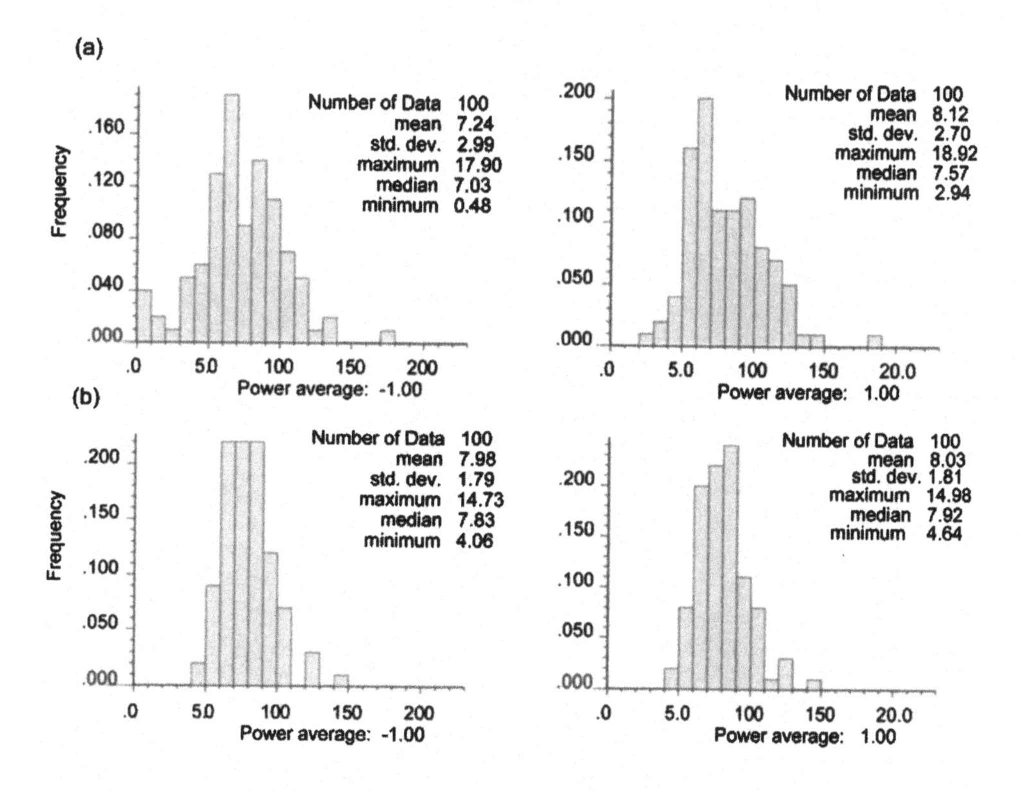

Figure 7.5 Histograms of block averaged (a) simulated field, and (b) kriged field from harmonic (power of −1.00) and arithmetic averaging (power of 1.00).

Remarks

Kriging under a multivariate Gaussian assumption provides local distributions of uncertainty. If we are only concerned with any one particular location, then backtransforming the distributions parameterized by the simple kriging mean and variance will be sufficient. However, in practice we often need to consider multiple locations simultaneously; the most common example is that of assessing resources/reserves for the area of interest. In these instances, the joint uncertainty is crucial for development decisions.

CHAPTER 8

Indicators

The indicator formalism is natural for categorical variables. Often, there is no intrinsic meaning or even ordering of numbers representing categorical variables. They are rock types, facies, or some other discrete description. The use of indicator variables facilitates numerical analysis of categorical variables. An indicator variable is defined for each category k at location $\mathbf{u}_\alpha$ as:

$$i(\mathbf{u}_\alpha;k) = \begin{cases} 1, & \text{if category k is present at location } \mathbf{u}_\alpha \\ 0, & \text{otherwise} \end{cases} \quad (8.1)$$

Where there are $k=1,...,K$ mutually exclusive and exhaustive categories; the single categorical variable at location $\mathbf{u}_\alpha$ is replaced with a vector of K indicator values. At the scale of the data, one indicator will be 1 and all others 0. Indicators for a categorical variable can be thought of as a probability coding of the data: the probability of the category prevailing at the specified location. Kriging and simulation of indicator variables permits inference of the probability and/or proportion of different categories (Journel, 1982; Journel, 1983; Journel, 1984). Estimated indicator values will be somewhere between 0 and 1.

The indicator approach has also been extended to continuous variables by considering a series of increasing thresholds z_k, $k=1,...,K$ that discretize the range of variability in the continuous variable (Journel, 1983):

$$i(\mathbf{u}_\alpha;z_k) = \begin{cases} 1, & \text{if } Z(\mathbf{u}_\alpha) \le z_k \\ 0, & \text{otherwise} \end{cases} \quad (8.2)$$

Solved Problems in Geostatistics. By O. Leuangthong, K.D. Khan, and C.V. Deutsch

Indicators for continuous variables may also be interpreted as a probability coding of the data value: the probability that the value is less than or equal to the specified threshold. The series of indicator values provides a discretized approximation of the cdf. The cdf at data locations is a step function, either 0 or 1. Kriging each binary indicator at unsampled locations will provide an estimate between 0 and 1, and this set of estimated proportions provides the distribution of uncertainty at each location. The choice of thresholds is important. Too few and there is a significant loss of resolution. Too many and parameterization of the indicator random variables becomes tedious.

The single categorical random variable or continuous random variable is replaced by an indicator vector random function of dimension K. This is required for a categorical random variable where the numerical categorical values rarely have intrinsic numerical meaning. This may be convenient for continuous variables that are not well characterized by a Gaussian distribution. Statistics of the indicator variables may be calculated. The mean and variance of the indictor variable are

$$\begin{aligned} E\{i(\mathbf{u}_\alpha;k)\} &= p_k \\ Var\{i(\mathbf{u}_\alpha;k)\} &= p_k(1-p_k) \end{aligned} \tag{8.3}$$

where p_k is the proportion of category k or the proportion of data less than the threshold $Z(\mathbf{u}_\alpha) \leq z_k$. The variance is a simple function of the mean because the indicator data are binary at the scale of the data.

A variogram model may be calculated and fitted for each category or threshold k=1,...,K. Indicator variograms are interpreted as the transition probability for that category/threshold and distance. Local estimates with kriging are interpreted as best estimates of the probability that the category prevails or that the continuous variable is less than the threshold. The estimated probability values must be consistent. They cannot be negative or >1, they must sum to 1 in the case of a categorical variable, and they must be nondecreasing in the case of a continuous variable. These fundamental properties must be honored because of the interpretation of the estimated indicators as probabilities. Often, post processing is required to ensure consistency of the estimated probabilities.

The local distributions can be retained and used directly as kriging estimates or they can be used in simulation.

This chapter aims to improve the reader's understanding of indicator transforms, indicator variograms and their interpretation, and finally indicator simulation for a categorical variable. The first problem demonstrates the relationship between the indicator variogram and the geometry of geological shapes. The second problem shows the analytical link between Gaussian variograms and indicator variograms. Finally, the third problem focuses on the steps relevant to performing an indicator simulation for categorical variables.

8.1 VARIOGRAM OF OBJECTS

Learning Objectives

The interpretation of the indicator variogram as a transition probability is best illustrated with a simple geometric example. This problem investigates indicator variograms using a simple object-based model that represents one geological rock type as an idealized object shape within another background rock type. The dependency of the indicator variogram on the size, shape, and proportion of the rock types is established theoretically. Although an analytical link is not always possible, it improves understanding in all cases.

Background and Assumptions

Object-based stochastic modeling allows the simulation of random processes with sharp boundaries and distinct shapes. In this problem, a simple object-based model is used to illustrate the nature of the indicator variogram.

The Bombing Model (Hall, 1988) is a Boolean model (Matheron, 1975) formed by placing random closed sets at the points of a Poisson process and taking the union of these sets (van Lieshout and van Zwet, 2000). An indicator variable is defined as: 1 for locations outside of the closed sets and 0 inside. The variogram for the 2D Bombing Model of circular objects can be calculated analytically as (Matern, 1960):

$$\gamma(\mathbf{h}) = p(1 - p^{Circ_a(|h|)}) \tag{8.4}$$

The variogram relation (8.4) has a form similar to the familiar indicator variance, that is, $Var\{I\} = p(1-p)$. The exponent term corresponds to the standardized (unit sill) hyperspherical variogram model for the dimension under consideration. In the case of 2D circles of diameter a, the exponent is the circular variogram parameterized by a:

$$Circ_a(|h|) = \begin{cases} \dfrac{2h}{\pi a}\sqrt{1-\left(\dfrac{h}{a}\right)^2} + \dfrac{2}{\pi}\arcsin\dfrac{h}{a}, & h \le a \\ 1, & h > a \end{cases}$$

In 1D, the exponent would be a variogram that is linearly increasing to the range a, then constant at 1. In 3D, the exponent is the well-known spherical variogram function (Srivastava, 1985; Deutsch, 1987).

Problem

(i) Write the expression for the variogram corresponding to the bombing model process in terms of spatial transition probabilities associated with the following indicator variable:

$$I(\mathbf{u}) = \begin{cases} 1, & \text{if } \mathbf{u} \text{ located outside an ellipsoid} \\ 0, & \text{otherwise} \end{cases} \tag{8.5}$$

For this derivation, consider 3D spheres of the same size positioned at random within a uniform background.

(ii) Plot the indicator variograms calculated and tabulated in Table 8.1 from the three bombing model realizations shown in Figure 8.1. Interpret the results and comment on the nature of the indicator variogram.

Table 8.1 Experimental Indicator Variogram Values Calculated on the Three Models Shown in Figure 8.1

lag	dist.	$\gamma_I(\mathbf{h})$ model 1	$\gamma_I(\mathbf{h})$ model 2	$\gamma_I(\mathbf{h})$ model 3
1	1.414	0.209	0.252	0.368
2	4.243	0.547	0.629	0.749
3	7.071	0.782	0.865	0.914
4	9.899	0.937	0.991	0.971
5	12.728	1.005	1.033	0.964
6	15.556	1.025	1.026	0.934
7	18.385	1.046	1.011	0.927
8	21.213	1.058	0.990	0.918
9	24.042	1.049	0.981	0.914
10	26.870	1.026	1.005	0.911
11	29.698	0.989	1.024	0.933
12	32.527	0.963	1.035	0.949
13	35.355	0.959	1.032	0.951
14	38.184	0.952	1.031	0.915
15	41.012	0.926	1.002	0.882

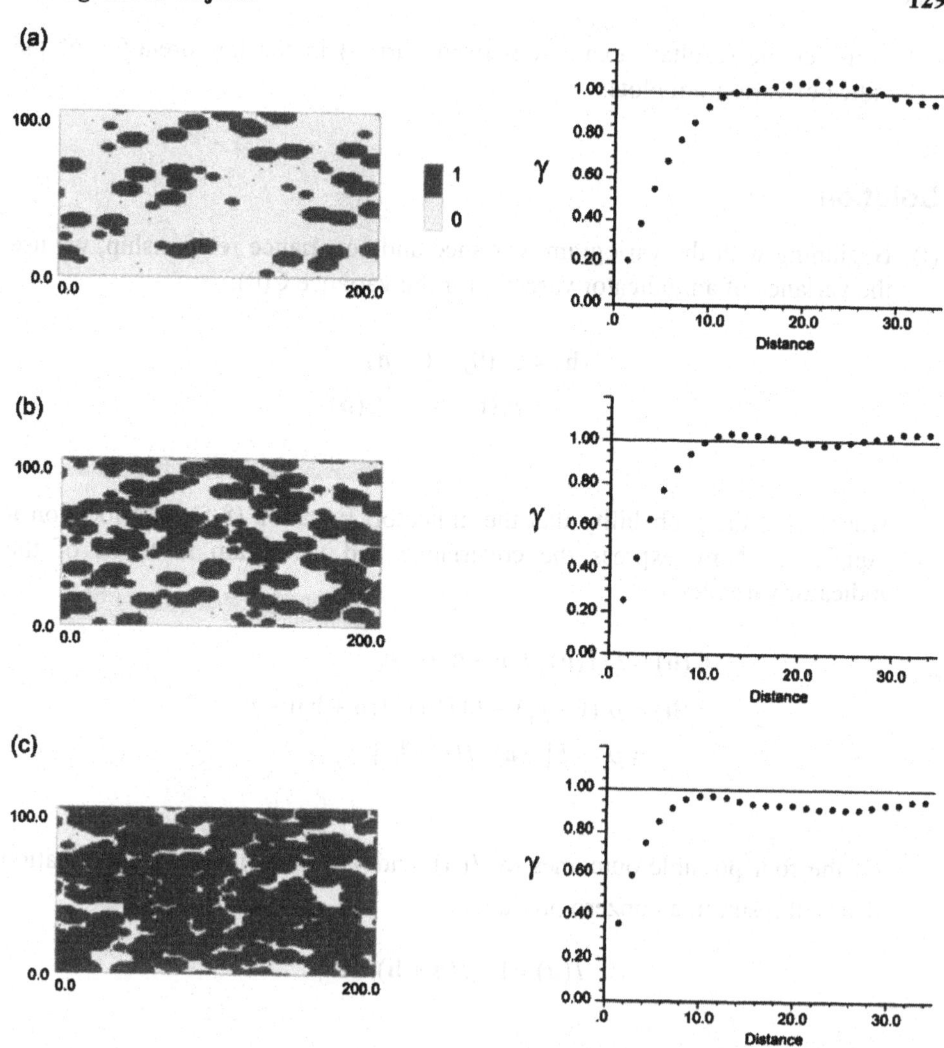

Figure 8.1 Three object-based models with ellipsoids distributed as a Poisson point process. The three models contain objects of three different sizes. Each model specifies a target proportion of ellipsoidal objects: (a) p=0.24, (b) p=0.54, (c) p=0.84.

Solution Plan

(i) Use the relationship between the variogram and covariance functions (see Chapter 5), the expression for the variance of an indicator variable [Equation (8.3)], and substitute the values for the possible outcomes of the indicator variogram to derive the expression for the indicator variogram in terms of indicator probabilities.

(ii) Consider the resultant expression from Part (i) in the interpretation of the indicator variogram plots.

Solution

(i) Beginning with the variogram, variance and covariance relationship, we use the variance of an indicator variable for the variance C(0):

$$\begin{aligned}\gamma_I(\mathbf{h}) &= C_I(0) - C_I(\mathbf{h}) \\ &= p_1(1-p_1) - C(\mathbf{h})\end{aligned}$$

where p_1 is the probability that the indicator, Equation (8.5), at a location $\mathbf{u}$ equals 1. Now express the covariance and variogram in terms of the indicator variable:

$$\begin{aligned}C(\mathbf{h}) &= E[I(\mathbf{u})\cdot I(\mathbf{u}+\mathbf{h})] - p_1^{\ 2} \\ \gamma(\mathbf{h}) &= p_1(1-p_1) - E[I(\mathbf{u})\cdot I(\mathbf{u}+\mathbf{h})] + p_1^{\ 2} \\ &= p_1 - E[I(\mathbf{u})\cdot I(\mathbf{u}+\mathbf{h})]\end{aligned}$$

Of the four possible outcomes of $I(\mathbf{u})$ and $I(\mathbf{u}+\mathbf{h})$, the only combination that will result in a nonzero product is

$$I(\mathbf{u}) = 1, \quad I(\mathbf{u}+\mathbf{h}) = 1$$

Therefore,

$$\gamma_I(\mathbf{h}) = p_1 - p_{11}(\mathbf{h}) \tag{8.6}$$

The frequency of the centers of spheres falling in a volume v follows a Poisson distribution with parameter av, where

$$p_1 = e^{-a\frac{4}{3}\pi r^3}$$

$$a = \frac{-\ln(p_1)}{\frac{4}{3}\pi r^3}$$

The probability of $I(\mathbf{u})=1 \cap I(\mathbf{u}+\mathbf{h})=1$ is

$$\begin{aligned} p_{11}(\mathbf{h}) &= \text{Prob}\{[I(\mathbf{u})=1 \mid I(\mathbf{u}+\mathbf{h})=1]\cdot \text{Prob}[I(\mathbf{u}+\mathbf{h})=1]\} \\ &= p_1 \cdot \text{Prob}\{I(\mathbf{u})=1 \mid I(\mathbf{u}+\mathbf{h})=1\} \end{aligned}$$

For $h > a$ $I(\mathbf{u})$ and $I(\mathbf{u}+\mathbf{h})$ are independent, and $p_{11}(\mathbf{h}) = p_1^{\,2}$. For $h \le a$

$$\text{Prob}\{I(\mathbf{u})=1 \mid I(\mathbf{u}+\mathbf{h})=1\} = e^{-av(h)}$$

where v(**h**) is the shaded volume shown in the following illustration:

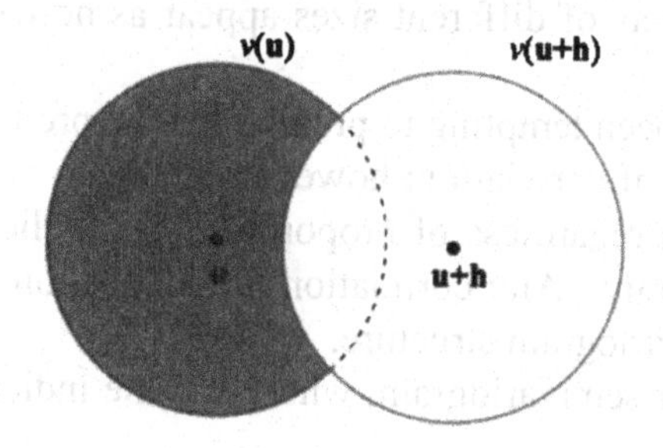

$$\begin{aligned} v(h) &= \pi r^2 h - \frac{\pi h^3}{12} \\ p_{11} &= p_1 \cdot e^{-av(h)} \\ &= p_1 \cdot e^{\left[\frac{\ln p_1}{\frac{4}{3}\pi r^3}\left(\pi r^2 h - \frac{\pi h^3}{12}\right)\right]} \\ &= p_1 \cdot p_1^{\frac{3}{4}\frac{h}{r} - \frac{1}{16}\frac{h^3}{r^3}} \\ &= p_1 \cdot p_1^{\frac{3}{2}\frac{h}{a} - \frac{1}{2}\left(\frac{h}{a}\right)^3} \\ &= p_1 \cdot p_1^{Sph(h)} \end{aligned}$$

where $Sph(\mathbf{h})$ is the spherical semivariogram function. Finally,

$$\begin{aligned}\gamma(\mathbf{h}) &= p_1 - p_1^2 = p_1(1-p_1) && \text{for } \mathbf{h} > a \\ &= p_1 - p_{11}(\mathbf{h}) && \text{for } \mathbf{h} \le a \\ &= p_1 - p_1 \cdot p_1^{Sph(\mathbf{h})} \\ &= p_1\left(1 - p_1^{Sph(\mathbf{h})}\right)\end{aligned}$$

(ii) Figure 8.1 is a plot of the semivariogram values given in Table 8.1 (at slightly higher sampling density using twice as many lags as provided) next to the corresponding three models. Despite the different proportion of objects embedded in each model, the indicator variogram range is effectively the same in each case. The objects are allowed to overlap in the models shown, resulting in amalgamated objects of different sizes in each model. Overlapping objects of different sizes appear as nested variogram structures (Figure 8.1).

It may have been tempting to predict that the proportion of objects would change the range of correlation; however, the range of correlation is constant at the object size regardless of proportion. This relies on the objects being positioned randomly. Any correlation in the location of the object centroids will change the variogram structure.

The indicator semivariogram, where k is the indicator category, or class, is calculated as:

$$\gamma_I(\mathbf{h}) = \frac{1}{2} E\{[I(\mathbf{u};k) - I(\mathbf{u}+\mathbf{h};k)]^2 \tag{8.7}$$

Considering pairs of data, there are four possible outcomes for the squared difference calculated in Equation (8.7):

$$\begin{aligned}&\text{(a) } I(\mathbf{u}) = 1, \quad I(\mathbf{u}+\mathbf{h}) = 1; \quad [I(\mathbf{u}) - I(\mathbf{u}+\mathbf{h})]^2 = 0 \\ &\text{(b) } I(\mathbf{u}) = 1, \quad I(\mathbf{u}+\mathbf{h}) = 0; \quad [I(\mathbf{u}) - I(\mathbf{u}+\mathbf{h})]^2 = 1 \\ &\text{(c) } I(\mathbf{u}) = 0, \quad I(\mathbf{u}+\mathbf{h}) = 1; \quad [I(\mathbf{u}) - I(\mathbf{u}+\mathbf{h})]^2 = 1 \\ &\text{(d) } I(\mathbf{u}) = 0, \quad I(\mathbf{u}+\mathbf{h}) = 0; \quad [I(\mathbf{u}) - I(\mathbf{u}+\mathbf{h})]^2 = 0\end{aligned}$$

Since only outcomes (b) and (c) contribute to the semivariogram, it is clear that the indicator variogram measures the frequency of transitions between indicator classes.

Although the images in Figure 8.1 appear quite different, the standardized indicator variograms have the same range and nearly the same shape. The range would be different in different directions because the objects are ellipsoidal and not spherical.

Remarks

There is a direct link between the geometry of objects and the indicator variogram. This provides an interesting link between indicator techniques and object-based modeling. For example, indicator variograms can be used to estimate the object size parameters needed in object-based modeling, which would be particularly useful when the objects amalgamate and individual object thicknesses are difficult to identify.

8.2 INDICATOR VARIOGRAMS AND THE GAUSSIAN DISTRIBUTION

Learning Objectives

Indicators are considered for continuous variables that are nonGaussian. Real data do not fit any congenial distribution, including the Gaussian one. An important question is how far the data deviate from the Gaussian model (Rossi and Posa, 1992). If the deviations are small, then it may be reasonable to use the simpler Gaussian approaches. Comparing actual indicator variograms with the ones theoretically predicted under a Gaussian assumption could often help determine the acceptability of the Gaussian model. Performing this check helps to understand the consequences of adopting the multi-Gaussian model.

Background and Assumptions

The multiGaussian assumption is often made without regard to how closely the data may actually follow a multivariate Gaussian distribution. Although the transformation of the original data to follow a normal distribution for the estimation and simulation is correct for the univariate histogram, geostatistical calculations are performed in multivariate Gaussian space. Therefore the results will inevitably have strong Gaussian characteristics, which may be inappropriate for some earth models (Journel and Deutsch, 1993; Zinn and Harvey, 2003). The convenience of the Gaussian model should not preclude checking the consequence of this stringent modeling assumption.

A statistically correct check for consistency with a multivariate Gaussian distribution would be impossible. Checking beyond second-order statistics is generally intractable due to the inability to infer the N-point statistics from the sample data (Goovaerts, 1997, p. 415). There are many second-order checks. One commonly advocated procedure (Deutsch and Journel, 1998) compares the actual variograms of the indicator transforms to that predicted by the multivariate Gaussian distribution. Consider a Gaussian variable, Y, and a specified threshold y_p:

$$\begin{aligned}\gamma_I(\mathbf{h}; y_p) &= C_I(0) - C_I(\mathbf{h}; y_p)\\ &= p(1-p) - \left(E\left[I(\mathbf{u}; p)\cdot I(\mathbf{u}+\mathbf{h}; p)\right] - p^2\right)\\ &= p - G(\mathbf{h}; y_p)\end{aligned} \tag{8.8}$$

where

$$G(\mathbf{h}; y_p) = \text{Prob}\{Y(\mathbf{u}) \le y_p, Y(\mathbf{u}+\mathbf{h}) \le y_p\} \tag{8.9}$$

The bivariate Gaussian cdf, $G(\mathbf{h}; y_p)$, is the noncentered indicator covariance for the threshold y_p. More details on relation (8.9) are available in (Journel and Posa, 1990; Goovearts, 1997). Y is the normal score transform of the original Z variable; however, the transformation of the data is not necessary to perform this check (Deutsch and Journel, 1998, p. 142). In that case, the indicator, $I(\mathbf{u}; p)$, is defined by the z_p cutoff on the same p-quantile value of the untransformed cdf.

Problem

Assess the consequences of adopting the multi-Gaussian assumption in modeling the spatial variability of gold grade at a specific site. Indicator variograms for specific y_p cutoffs have been calculated and tabulated (Table 8.2) along with the theoretical bivariate Gaussian values for comparison (Table 8.3).

Solution Plan

Compare the variogram values at specific lags between the experimental indicator-transform defined on the y_p, p=1,...,5 cutoffs and the corresponding theoretical values for a truly bivariate normal attribute.

Table 8.2 Experimental Indicator Semivariogram Values for Five Indicator Classes Defined from the cdf of Gold Grade

h	$\gamma(\mathbf{h}; y_{0.1})$	$\gamma(\mathbf{h}; y_{0.25})$	$\gamma(\mathbf{h}; y_{0.5})$	$\gamma(\mathbf{h}; y_{0.75})$	$\gamma(\mathbf{h}; y_{0.9})$
0	0.000	0.000	0.000	0.000	0.000
6	0.253	0.250	0.297	0.437	0.475
30	0.425	0.467	0.556	0.694	0.781
60	0.599	0.634	0.703	0.813	0.869
90	0.710	0.765	0.864	0.974	1.031
120	0.846	0.899	0.993	1.077	1.114
150	0.981	1.007	1.043	1.051	1.051
180	1.084	1.103	1.115	1.084	1.063
210	1.214	1.202	1.158	1.080	1.032
240	1.292	1.240	1.144	1.024	0.974
270	1.283	1.203	1.065	0.929	0.884

Table 8.3 Theoretical Bivariate Gaussian Indicator Semivariogram Values for Comparison with the Indicator Variograms in Table 8.2.*

h	$\gamma(\mathbf{h}; y_{0.1})$	$\gamma(\mathbf{h}; y_{0.25})$	$\gamma(\mathbf{h}; y_{0.5})$
0	0	0	0
6	636	0.574	0.549
30	0.66	0.599	0.573
60	0.784	0.727	0.702
90	0.873	0.828	0.808
120	0.935	0.907	0.894
150	0.975	0.962	0.956
180	0.996	0.994	0.993
210	1	1	1
240	1	1	1
270	1	1	1

* Note that the theoretical indicator variograms are symmetrical about the p=0.5 threshold, so $\gamma(\mathbf{h}; y_{0.1})$ is equivalent to $\gamma(\mathbf{h}; y_{0.9})$, for example.

Solution

In this case, the gold grades are more continuous at lower values [i.e., $p \leq 0.1$, Figure 8.2(a)] than predicted by the bivariate Gaussian model. The experimental variogram has a significantly lower nugget. The median indicator [Figure 8.2(b)] is fit reasonably well by the Gaussian model, but the continuity at longer range is exaggerated. High values are often more discontinuous [Figure 8.2(c)], having a higher nugget effect, and rising to the variogram sill faster with increasing lag distance than predicted by the bivariate Gaussian model.

The consequence of modeling the gold grade at this site under the multi-Gaussian assumption is that low grade zones will be more disconnected than the data suggests, and high grades will have exaggerated connectivity at moderate-to-large lag distances.

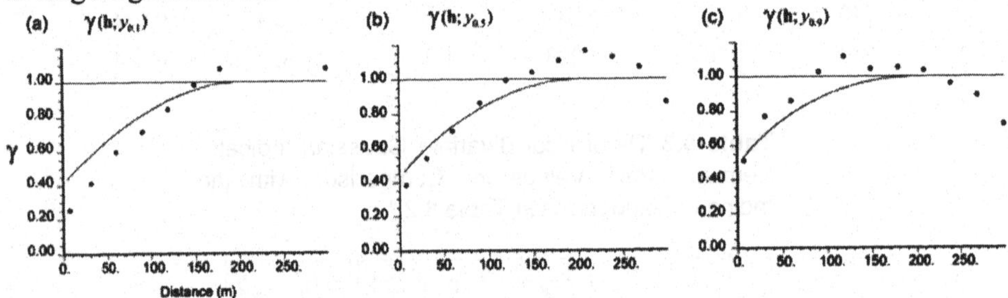

Figure 8.2 Data from Table 8.3 for indicators defined on (a) p=0.1, (b) p=0.5, and (c) p=0.9 plotted as points with theoretical Gaussian indicator variograms plotted as curves.

Remarks

In problem settings where clustered or connected spatial patterns of extreme values are important, such as flow and transport models, the multi-Gaussian assumption is often questionable (Gomez-Hernandez and Wen, 1998); however, the indicator formalism requires more data to infer all of the required statistics. The Gaussian distribution is widely used because of its simplicity. A common approach to overcome the limitations of the Gaussian distribution is to first create models of rock types that capture large scale and complex features of connectivity; then, Gaussian techniques are used within each rock type.

8.3 INDICATOR SIMULATION FOR CATEGORICAL DATA

Learning Objectives

Indicator formalism provides a means to determine a nonparametric conditional distribution. This problem demonstrates how indicator kriging is used to construct local distributions of uncertainty, from which we can draw simulated values via Monte Carlo simulation. While this form of simulation is common for both categorical and continuous variables, the focus of this problem is categorical data.

Background and Assumptions

Indicator kriging is a powerful method for constructing local distributions of uncertainty. The ccdfs built by indicator kriging are considered nonparametric in that they are not simply defined by a few parameters, such as the mean and variance at a location; rather, they are built from cumulative proportions (i.e., by counting) of the indicators prevailing at each location (Isaaks and Srivastava, 1989, pp. 421–424).

Recall the definition of an indicator variable for categorical data [see Equation (8.1)] and its first two moments [Equation (8.3)]. Further, categorical indicators are mutually exclusive; $i(\mathbf{u};k)\cdot i(\mathbf{u};k')=0,\ \forall\, k\neq k'$, and exhaustive such that $\sum_{k=1}^{K} i(\mathbf{u};k)=1$. Simple kriging of this indicator variable is referred to as indicator kriging and yields an estimate of this form:

$$\left[i(\mathbf{u};k)\right]_{SK}^{*}=\sum_{\alpha=1}^{n}\lambda_{\alpha}(\mathbf{u}_{\alpha};k)\,i(\mathbf{u}_{\alpha};k)+\left[1-\sum_{\alpha=1}^{n}\lambda_{\alpha}(\mathbf{u}_{\alpha};k)\right]p_{k} \qquad (8.10)$$

where p_k is the stationary mean of the indicator, which is its global proportion (pdf). Unlike kriging of continuous variables, the kriging of binary values yields an estimate that is continuous. In fact, the indicator kriged estimate is the conditional expectation $E\{i(\mathbf{u};k)\mid(n)\}$, where n data conditioning events occur in the search neighborhood. Thus, indicator kriging of each category yields the local proportion of the category. Note that the IK of each category then requires an indicator variogram for each category k. These locally conditioned proportions can then be used to construct a local distribution of uncertainty about the categorical RV, $I(\mathbf{u};k)$. Sequential simulation from this ccdf is known as sequential indicator simulation.

Order relations problems are common when the distribution of uncertainty is obtained using this nonparametric approach. The ccdfs are constructed by kriging each category separately from one another, therefore, it is inevitable that these "independent" estimation fields will sometimes produce discrete ccdf values whose sum does not equal 1.0. In the categorical case, a simple restandardization of the ccdf can resolve this issue. For continuous variables, another instance of order relations problems occurs when the ccdf does not increase monotonically; one proposed solution is to apply and average upward and downward corrections to the ccdf (Deutsch and Journel, 1998, pp.81–86). This latter issue does not present a problem in the categorical case since the ordering of the categories in the cdf is inconsequential.

Problem

A field of interest has three main rock types and it is necessary to construct a model of spatial uncertainty characterizing the distribution and occurrence of the rock types. There are five data retained within the local neighborhood of a single unsampled location, **u** (Figure 8.3; Table 8.4).

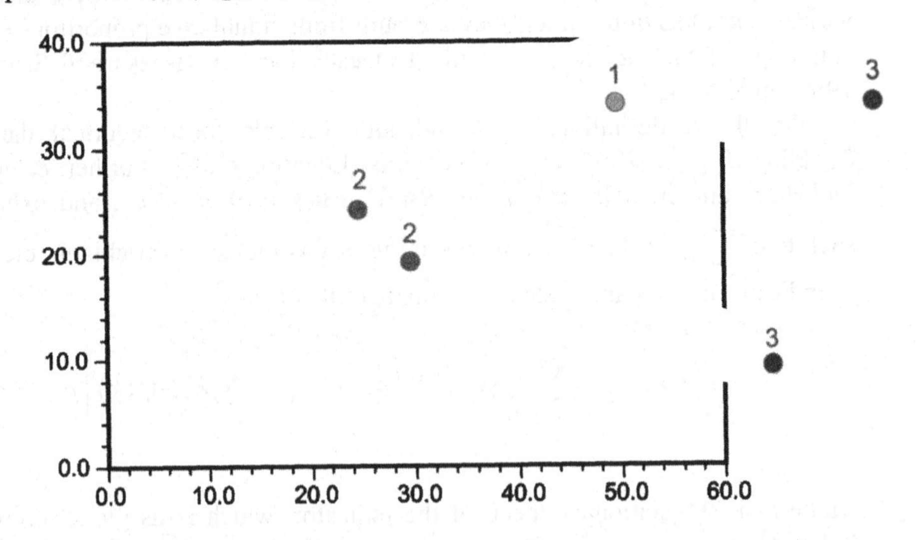

Figure 8.3 Location map of data with rock type labeled, and unsampled location, **u**.

Table 8.4 Data Coordinates and Rock Type

ID	X	Y	Rock type
1	5	20	2
2	10	15	2
3	45	5	3
4	55	30	3
5	30	30	1
u	35	15	

(i) Build the local conditional probability distribution by modeling the probability of encountering any of the three rock types at location **u** via indicator kriging. Plot the resultant conditional probability density and cumulative distribution function.

(ii) Simulate 100 realizations from the local cumulative distribution of rock types and plot the histogram of the simulated values.

The rock types are coded 1–3 for convenience. The global proportions or prior probabilities of each rock type are p_1=0.3; p_2=0.5; p_3=0.2. Use the following indicator variogram model for all three rock types:

$$\gamma_k(\mathbf{h}) = \sigma_k^2 \left(0.4Sph_{a=45}(\mathbf{h}) + 0.6Sph_{a=220}(\mathbf{h})\right) \tag{8.11}$$

where the sill of the variogram model, σ_k^2, is a function of the global mean, p_k; $k=1,\ldots,3$ of the indicator.

Solution Plan

(i) Transform the rock type data to three binary indicator variables of the type in Equation (8.1). Set up the simple indicator kriging system of equations for each indicator variable (see Problem 6.1) using the indicator variogram model provided in Equation (8.11). Solve for the vector of kriging weights obtained from each of the three kriging systems.

The simple kriging estimate corresponding to the estimated proportion of rock type k at the unsampled location is given by Equation (8.10). Compute the required SK estimates designating the probability of encountering each of the three rock types. Plot the three SK estimates as a pdf and a cdf and ensure that the conditional cumulative distribution is valid.

(ii) Draw 100 random numbers in [0,1] and the corresponding rock type realization at location **u**. Plot the cdf of the simulated realizations at location **u**.

Solution

(i) For each category, the indicator transform for the discrete variable is

$$I_k(\mathbf{u}) = \begin{cases} 1, & \text{if } \mathbf{u} \text{ is in category } k \\ 0, & \text{otherwise} \end{cases}, \quad k = 1,2,3$$

The system of SK equations for each indicator is set up as:

$$\begin{bmatrix} C_{1-1} & C_{1-2} & C_{1-3} & C_{1-4} & C_{1-5} \\ C_{2-1} & C_{2-2} & C_{2-3} & C_{2-4} & C_{2-5} \\ C_{3-1} & C_{3-2} & C_{3-3} & C_{3-3} & C_{3-3} \\ C_{4-1} & C_{4-2} & C_{4-3} & C_{4-4} & C_{4-5} \\ C_{5-1} & C_{5-2} & C_{5-3} & C_{5-4} & C_{5-5} \end{bmatrix} \begin{bmatrix} \lambda_1 \\ \lambda_2 \\ \lambda_3 \\ \lambda_4 \\ \lambda_5 \end{bmatrix} = \begin{bmatrix} C_{1-\mathbf{u}} \\ C_{2-\mathbf{u}} \\ C_{3-\mathbf{u}} \\ C_{4-\mathbf{u}} \\ C_{5-\mathbf{u}} \end{bmatrix}$$

which amounts to the following (standardized covariance) system for the data configuration in Figure 8.3 and the given indicator variogram:

$$\begin{bmatrix} 1.000 & 0.878 & 0.429 & 0.395 & 0.574 \\ 0.878 & 1.000 & 0.473 & 0.409 & 0.599 \\ 0.429 & 0.473 & 1.000 & 0.574 & 0.547 \\ 0.395 & 0.409 & 0.574 & 1.000 & 0.599 \\ 0.574 & 0.599 & 0.547 & 0.599 & 1.000 \end{bmatrix} \begin{bmatrix} \lambda_1 \\ \lambda_2 \\ \lambda_3 \\ \lambda_4 \\ \lambda_5 \end{bmatrix} = \begin{bmatrix} 0.533 \\ 0.599 \\ 0.760 \\ 0.599 \\ 0.733 \end{bmatrix}$$

Inverting the LHS data-to-data covariance matrix and multiplying by the RHS covariances yields the kriging weights:

$$\begin{bmatrix} \lambda_1 & \lambda_2 & \lambda_3 & \lambda_4 & \lambda_5 \end{bmatrix}^T = \begin{bmatrix} -0.085 & 0.211 & 0.460 & 0.063 & 0.366 \end{bmatrix}$$

Using these weights, the IK estimates for each rock type category [see Equation (8.10)] are

$$i_1^*(\mathbf{u}) = 0.362$$
$$i_2^*(\mathbf{u}) = 0.118$$
$$i_3^*(\mathbf{u}) = 0.520$$

which are plotted as the conditional pdf and cdf in Figure 8.4.

(ii) Drawing 100 realizations from the cdf in Figure 8.4 yields 100 realizations of the rock type category at location $\mathbf{u}$. A distribution of uncertainty constructed for these realizations should look similar to the distributions of Figure 8.4.

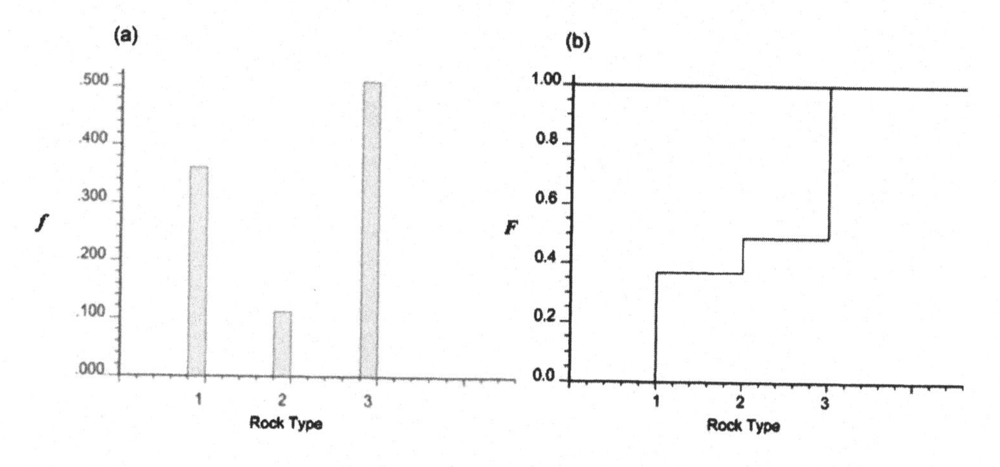

Figure 8.4 Conditional (a) pdf and (b) cdf derived from indicator kriging of the rock type data

Remarks

No order relation issues are encountered in this problem because the same indicator variogram was used for all categories and no negative weights were encountered. In practice, restandardization is commonly required because kriged estimates are calculated independently and different categories require different variogram models.

Recall that unlike continuous variables, ordering for categorical data is often unimportant. Consequently, the second type of order relations problems discussed above that deal with a nonmonotonic increase of estimated proportions is only an issue when indicator simulation is used for continuous variables.

Note also that the estimation variance for indicators is never called upon in the construction of the distribution of uncertainty. Local distributions are constructed directly from the kriged estimates of the categories or thresholds.

CHAPTER 9

Multiple Variables

Most geostatistical modeling studies consider two or more variables. There are often multiple elements of interest: contaminants, rate constants, and other rock properties. These variables are often related; it is rare for multiple variables from a particular geological site to be independent. This chapter deals with some theory and practice of estimation and simulation of multiple variables. More detailed theory on this subject can be found in Chilès and Delfiner (1999), Daly and Verly (1994), Goovearts (1997), Goulard (1989), Goulard and Voltz (1992), Hoef and Cressie (1993), Johnson (1987), Rivoirard (2001), Vargas-Guzman et al. (1999), Wackernagel et al. (1989) and Wackernagel (1994, 2003).

The variogram and covariance are easily extended to multiple variables. In its conventional form, the variogram measures the variability between two data events: the same variable separated by a lag vector **h**. Now, we must consider both variables at one location and both variables separated by **h**. The same principles of variogram calculation apply, only there are more variograms to calculate and fit. Consider M regionalized variables. As illustrated below, there are M direct variograms and $M(M-1)$ cross variograms. The cross variogram between variable i and j is the same as between j and i; therefore, there are only $M(M-1)/2$ cross variograms required in practice.

These direct and cross variograms cannot be considered independently. Cross correlation between two variables entails that the cross variograms must share some features. The higher the correlation between the two variables, the more similar the two direct variograms must be. Taken all together, the M direct variograms and $M(M-1)/2$ cross variograms must form a valid model of coregionalization for subsequent kriging and simulation. This represents a lot of work in practice; therefore, a number of practical shortcuts have been developed.

Solved Problems in Geostatistics. By O. Leuangthong, K.D. Khan, and C.V. Deutsch

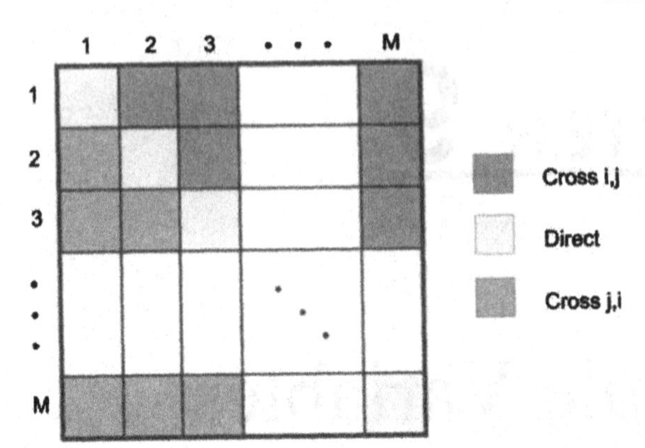

The first problem lays the groundwork for estimation/simulation of multiple variables by addressing the issue of simultaneously modeling multiple direct and cross variograms. The second problem looks at the most common types of cosimulation used in practice, namely, Gaussian cosimulation. The third problem explores the issue of estimation using different data of varying support volumes.

9.1 LINEAR MODEL OF COREGIONALIZATION

Learning Objectives

The linear model of coregionalization (LMC) is the only practical approach to simultaneously model direct and cross variograms in a multivariate setting. An LMC is suitable in cokriging and cosimulation. This problem demonstrates what is required when building a linear model of coregionalization for cokriging.

Background and Assumptions

A regionalized variable is described by local observations and statistical parameters that permit inference at unsampled locations. A univariate distribution and a variogram describe the regionalized variable, that is, they define a model of regionalization. In the presence of multiple variables and covariables, we must define a model of coregionalization.

For the multivariate case of M variables, the model of coregionalization is the set of M x M direct (or auto) and cross variograms, $2\gamma_{i,j}(\mathbf{h})$, $i,j=1,...,M$. These variograms must be simultaneously fit to be negative definite (or the covariances must be positive definite), that is, physically plausible (Goulard and Voltz, 1992). The linear model of coregionalization assumes that all M variables are a linear combination of L underlying independent factors: $l=1,...,L$. The weighting of factors in the different variables controls the relationship between these variables. Two variables that give more weight to a particular factor will be positively correlated. Two variables that give a particular factor weight with a different sign will be negatively correlated. It is not necessary to calculate the underlying independent factors. In practice, it would be impossible to unambiguously determine such factors; however, if the variables were a linear combination of independent factors, all direct and cross variograms would be a linear combination of the variograms of the underlying factors:

$$\gamma_{ij}(\mathbf{h}) = \sum_{l=0}^{L} b^l_{ij} \Gamma^l(\mathbf{h}) \tag{9.1}$$

The basic models, $\Gamma^l(\mathbf{h})$, $l = 0, \ldots, 1$, correspond to well-known valid variogram models standardized to a unit sill. By convention, the $l=0$ variogram is a pure nugget. The remaining variograms are independent of one another. The ranges of correlation and anisotropy are adjusted to fit the experimental variograms.

The practice of fitting a variogram model as a linear combination of elementary nested structures or models (spherical, exponential, etc.) is called building a model of regionalization (Journel and Huijbregts, 1978). Constructing a variogram model for a single variable is really a special case of modeling a (co) regionalization between multiple variables (Goovaerts, 1997).

For the multivariate case of M variables, the LMC is the set of M x M auto and cross semivariograms. Building a LMC requires choosing the family of variogram models $\Gamma^l(\mathbf{h}), l = 0,\ldots,L$ with the $(L+1)\cdot M^2$ parameters b^l_{ij}. This requires the simultaneous fitting of $M(M+1)/2$ permissible variogram models $\gamma_{ij}(\mathbf{h}), i, j = 1,\ldots,M$. The variance contribution parameters, b^l_{ij}, must be selected so as to ensure the $(L+1)$ matrices of coefficients b^l_{ij} are all positive semi-definite (Journel and Huijbregts, 1978; Isaaks and Srivastava, 1989; Goovearts, 1997).

Often, variables are considered two-by-two. For two variables, the above-stated condition for fitting a LMC requires that the following constraints are met

$$\begin{aligned} b_{ii} &\geq 0 \\ b_{jj} &\geq 0 \\ b^l_{ii} \cdot b^l_{jj} &\geq b^l_{ij} \cdot b^l_{ji}, \forall i, j, l \end{aligned} \tag{9.2}$$

Automatic fitting is almost always performed when $M>2$; the positive semi-definiteness of the coefficients is enforced by the program. The positive semi-definiteness of the coefficient matrices can be checked by calculating the eigenvalues. All of them must be nonnegative (Goovaerts, 1994).

Problem

Consider an oil sands data set consisting of two variables (1) the fraction of bitumen, and the (2) fraction of fine-grained sediment in the samples within a particular geological layer. The cross plot of the normal scores of these variables is shown in Figure 9.1. The direct variograms for bitumen and fine content and the corresponding cross variogram are given in Figure 9.2. These variograms may be fit by L=3 spherical structures with isotropic ranges of a_1=200 m, a_2 = 500 m, and a_3 = 1000 m. Write out a valid LMC given this information.

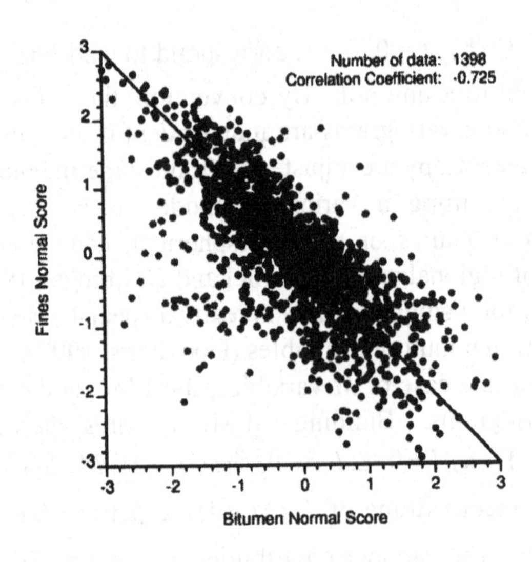

Figure 9.1 Cross plot of bitumen and fines content (normal scores units).

Solution Plan

The range parameters of the three fitted variograms are given, so all that remains to be determined is a set of valid coefficients b^l_{ij} subject to the constraints in Equation 9.2. Choose variance/covariance contributions, b, to fit each of the variograms and cross variogram. Verify that the above constraints are satisfied. This process is generally iterative. The data files are provided (*9.1_cv_bit.var*, *9.1_cv_fines.var*, and *9.1_cv_cross.var*).

Fitting a LMC without the aid of fully automatic software can be difficult. This problem is highly constrained and straightforward. In practice, it is best to fit the cross variogram(s) first because any structure in the cross variogram between variable i and j must also appear in the direct variograms of variables i

and j. Variogram structures that appear on a direct variogram need not necessarily appear on any other direct variogram or cross variogram. Also, note that the direct variogram models of two variables must be very similar if the cross correlation between the two variables is strong (positive or negative).

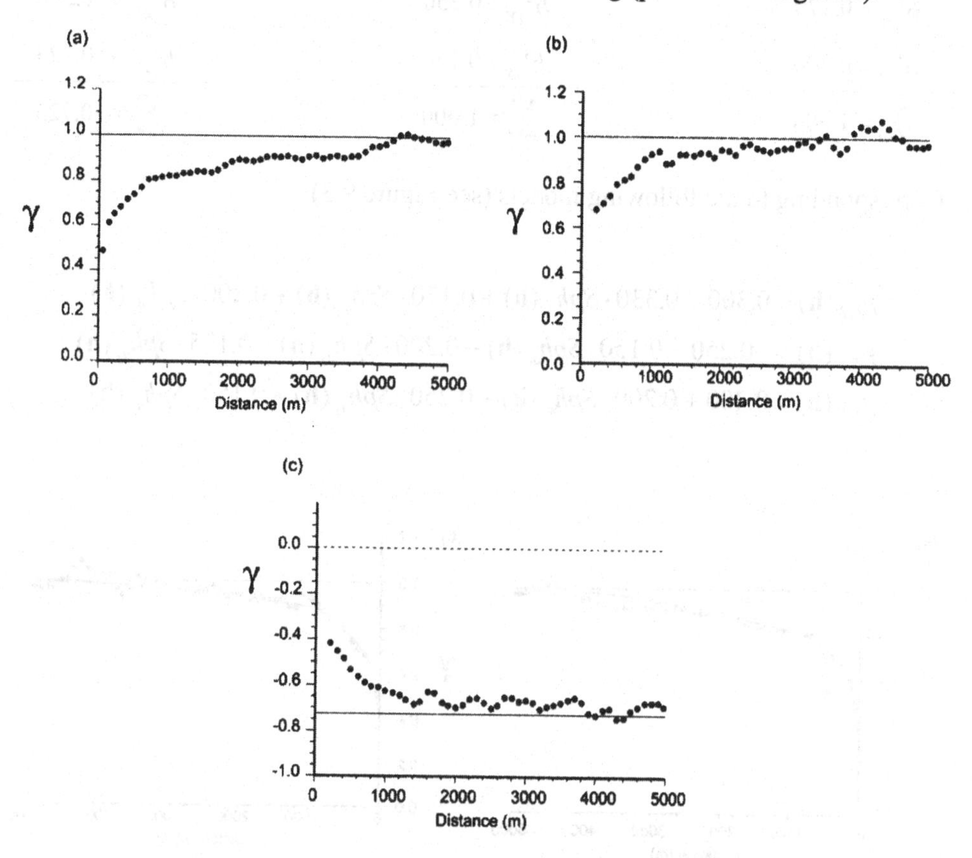

Figure 9.2 Experimental variograms for (a) bitumen content, (b) fines content, (c) cross semivariogram between bitumen and fines.

A set of b values must be chosen so that the fit appears reasonable and the sills of the direct and cross variograms is correct. In the case of standardized variables, the sill of the direct variograms must be 1 and the sill of the cross variograms must be the cross correlation between collocated values of those variables.

Solution

Below is a possible set of variance/covariance coefficients that yields a valid LMC:

$b^0_{ZZ} = 0.300$	$b^0_{YY} = 0.400$	$b^0_{ZY} = -0.250$
$b^1_{ZZ} = 0.330$	$b^1_{YY} = 0.200$	$b^1_{ZY} = -0.150$
$b^2_{ZZ} = 0.170$	$b^2_{YY} = 0.250$	$b^2_{ZY} = -0.200$
$b^3_{ZZ} = 0.200$	$b^3_{YY} = 0.150$	$b^3_{ZY} = -0.125$
$\sum = 1.000$	$\sum = 1.000$	$\sum = -0.725$

Corresponding to the following models (see Figure 9.3):

$$\gamma_{Z,Z}(\mathbf{h}) = 0.300 + 0.330 \cdot Sph_{a_1}(\mathbf{h}) + 0.170 \cdot Sph_{a_2}(\mathbf{h}) + 0.200 \cdot Sph_{a_3}(\mathbf{h})$$
$$\gamma_{Z,Y}(\mathbf{h}) = -0.250 - 0.150 \cdot Sph_{a_1}(\mathbf{h}) - 0.200 \cdot Sph_{a_2}(\mathbf{h}) - 0.125 \cdot Sph_{a_3}(\mathbf{h})$$
$$\gamma_{Y,Y}(\mathbf{h}) = 0.400 + 0.200 \cdot Sph_{a_1}(\mathbf{h}) + 0.250 \cdot Sph_{a_2}(\mathbf{h}) + 0.150 \cdot Sph_{a_3}(\mathbf{h})$$

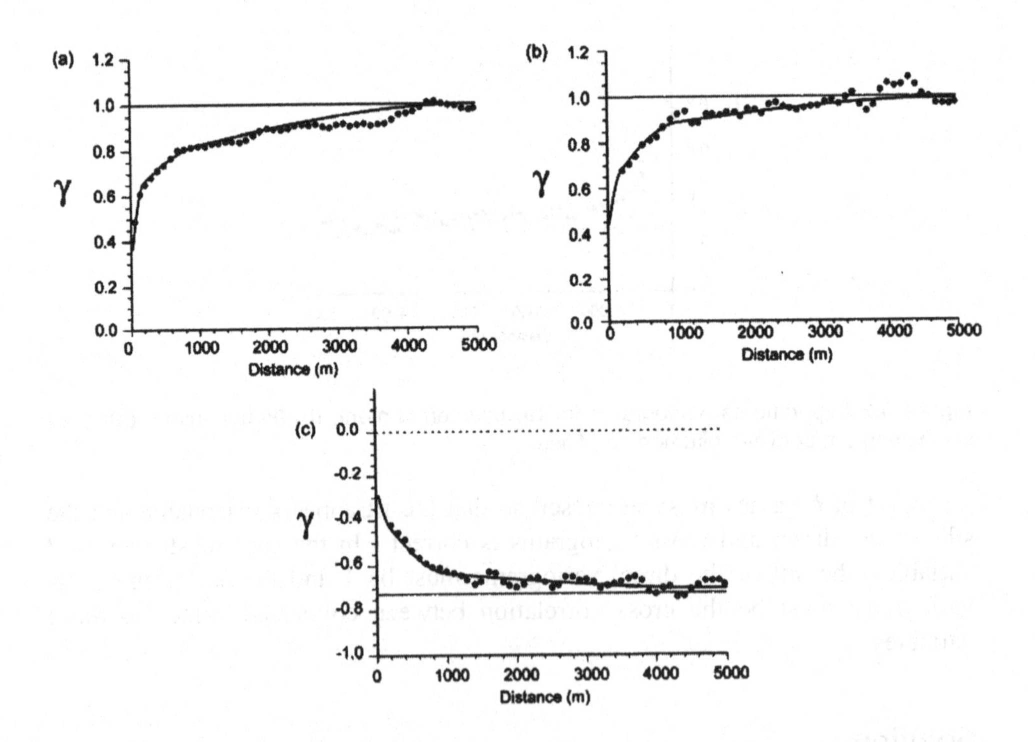

Figure 9.3 Experimental semivariograms with fitted curves for (a) bitumen content, (b) fines content, and (c) cross semivariogram between bitumen and fines.

Remarks

It is reasonably straightforward to fit an LMC for two variables. As the number of variables increases, the number of direct and cross variograms to fit simultaneously quickly becomes a very difficult problem. With four variables, a total of 10 variograms must be fitted simultaneously. Given that most natural attributes exhibit an anisotropic continuity structure, we would require fitting these 10 variograms in three directions. While automatic fitting could be used, it is fairly standard practice to consider a simpler alternative to the LMC model, such as *collocated* cokriging with a Markov model (Almeida and Journel, 1994), which is covered in the next problem.

9.2 GAUSSIAN COSIMULATION

Learning Objectives

Understand the common practice of collocated cokriging in Gaussian simulation. The linear model of coregionalization is often considered to be unworkable even in the presence of two variables. This problem shows why the simplicity of the Markov model often makes it a widely used cosimulation method available in many commercial and public domain software packages.

Background and Assumptions

We are almost always interested in multiple secondary data and/or multiple variables of interest for geostatistical modeling. Relationships between multiple variables can be complex; however, the correlation coefficient is a widely used statistic that succinctly summarizes how two variables are related. This parameter is the only one needed for the bivariate Gaussian distribution between two standard normal variables. The spatial cross correlation may be possible to infer where there are enough data of all types, and this leads to fitting a LMC (see Problem 9.1) and cokriging.

The cokriging estimator accounting for two secondary variables has the form:

$$z^*(\mathbf{u}) = \sum_{\alpha=1}^{n_z} \lambda_\alpha z(\mathbf{u}_\alpha) + \sum_{\alpha=1}^{n_{y1}} \lambda'_\alpha y_1(\mathbf{u}'_\alpha) + \sum_{\alpha=1}^{n_{y2}} \lambda''_\alpha y_2(\mathbf{u}''_\alpha) \quad (9.3)$$

where Z is a primary attribute of interest and Y_1 and Y_2 are two different secondary attributes related to Z. It is possible that all three variables are sampled at the same locations resulting in a *homotopic* multivariate sample set. Perhaps in practice, we more often see *heterotopic* data sets, where there are multiple secondary data sampled at different locations with different data densities (Myers, 1991). For example, in addition to the primary attribute, Z, there may be a secondary data, Y_1, that is exhaustively sampled and another data type, Y_2, with sparse sampling. The exhaustively sampled data could come from some form of remote sensing, such as satellite imaging or seismic data (Doyen, 1988).

The burden of modeling the cross-correlation needed for a full model of coregionalization prompted the popular collocated cokriging approach originally proposed by Xu et al. (1992) and developed by Almeida (1993). There are two characteristic features of this technique: (1) only the secondary data at the location being estimated are used, and (2) the cross correlation function is assumed proportional to the correlation function of the primary variable.

Secondary data are often densely sampled and relatively smoothly varying. For example, this is the case with remote sensing data sets that have been processed through one or more filters. In this case, the redundancy between nearby data may be so severe that only the collocated secondary datum is truly relevant. A reasonable approximation may be obtained by limiting the use of secondary data to the collocated value. This first choice removes the need for the variogram or correlogram function of the secondary data. The variogram of the primary variable and the cross variogram between the primary and all secondary variables is still required.

The second characteristic feature of collocated cokriging is to assume that the cross correlation function is proportional to the correlation function of the primary variable. This assumption is valid under a Markov model, that is, the collocated secondary data perfectly screens all other secondary data (Shmaryan and Journel, 1999; Journel, 1999b). This leads to the following relation for the cross correlation function:

$$\rho_{12}(\mathbf{h}) \simeq \rho_{12}(0)\rho_{11}(\mathbf{h}) \tag{9.4}$$

where subscripts denote the pairing of the primary (1) and secondary (2) variables, and $\rho(\mathbf{h})$ is the covariance function, $C(\mathbf{h})$, standardized by the corresponding product standard deviations. The simple scaling of the primary correlogram in Equation 9.4 by the correlation coefficient $\rho_{12}(0)$ between the primary and secondary data is the reason for adopting the Markov model.

An estimate of the parameters of the local ccdfs can be obtained using either full cokriging with a LMC or collocated cokriging assuming the Markov model of coregionalization. Simulating from these distributions to generate multiple realizations conditioned on all of the input data and preserving the spatial correlation and cross-correlation between the different data types results in a complete model of joint variability consistent with all of the data.

The methodological difference between the two approaches is in the hierarchical construction of the ccdfs. In the following problem, we are interested in estimating and simulating the value of two primary attributes Z_1 and Z_2 in the presence of another exhaustively sampled secondary variable Y_1. Full cokriging allows an essentially parallel computation of the two ccdfs where the kriging equations are solved with the same LHS data-to-data covariance matrix, but different RHS covariance vectors. Collocated cokriging requires a sequential or hierarchical construction. For example, one primary attribute, say Z_1, is conditioned on the collocated secondary variable Y_1; and then the second primary variable, Z_2, is conditioned on both the collocated secondary variable Y_1 and the simulated value of Z_1 at the same location.

Problem

Consider two primary variables Z_1 and Z_2 that are homotopically sampled. A secondary, fully exhaustive variable, Y, is available. All variables are standard normal. We are interested in simulating Z_1 and Z_2 at location $\mathbf{u}_0$, given the data configuration in Figure 9.4. For this configuration, generate 100 realizations for Z_1 and Z_2 by

(i) Full cosimulation using simple cokriging with the following LMC:

$$\begin{aligned}
\gamma_{Z1}(\mathbf{h}) &= 0.5 \cdot Sph_{a=120}(\mathbf{h}) + 0.5 \cdot Sph_{a=800}(\mathbf{h}) \\
\gamma_{Z2}(\mathbf{h}) &= 0.2 \cdot Sph_{a=120}(\mathbf{h}) + 0.8 \cdot Sph_{a=800}(\mathbf{h}) \\
\gamma_{Y}(\mathbf{h}) &= 0.8 \cdot Sph_{a=120}(\mathbf{h}) + 0.2 \cdot Sph_{a=800}(\mathbf{h}) \\
\gamma_{Z1-Z2}(\mathbf{h}) &= -0.2 \cdot Sph_{a=120}(\mathbf{h}) - 0.3 \cdot Sph_{a=800}(\mathbf{h}) \\
\gamma_{Z1-Y}(\mathbf{h}) &= -0.3 \cdot Sph_{a=120}(\mathbf{h}) - 0.1 \cdot Sph_{a=800}(\mathbf{h}) \\
\gamma_{Z2-Y}(\mathbf{h}) &= 0.3 \cdot Sph_{a=120}(\mathbf{h}) + 0.3 \cdot Sph_{a=800}(\mathbf{h})
\end{aligned} \tag{9.5}$$

(ii) Collocated cosimulation of Z_1 using Y, and collocated cosimulation of Z_2 conditional to Z_1 and Y. In both cases, use simple cokriging.

An important aspect of the problem is to compare the results to see how close the collocated option comes to full cokriging with a LMC.

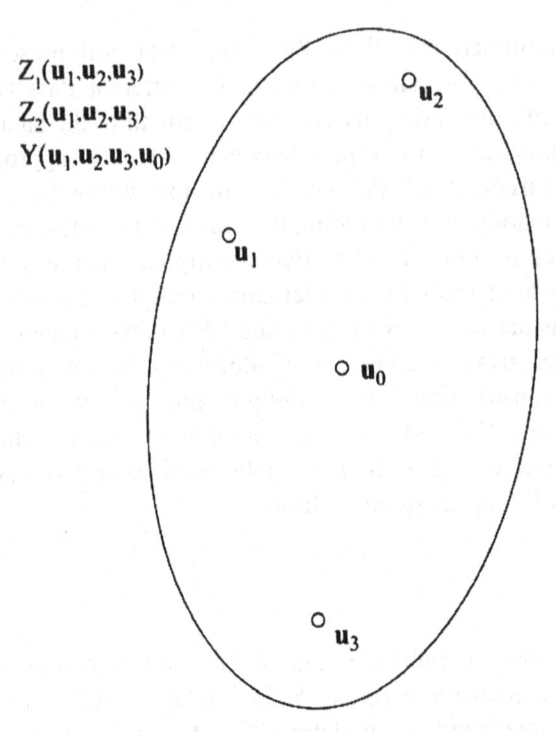

Figure 9.4 Location map of homotopic multivariate data. Data coordinates given in Table 9.1.

Table 9.1 Data Location Coordinates and Normal-Score Values of Available Data

Data	X (m)	Y (m)	Y	Z_1	Z_2
u_1	20	160	−0.6	0.5	−0.2
u_2	80	220	−1.2	0.9	0.1
u_3	50	30	1.4	−1.9	1.1
u_0	55	110	1		

Solution Plan

(i) Set up the data-to-data covariance matrix that comprises the LHS of the cokriging system. This matrix consists of n_i x n_j symmetric submatrices along the diagonal for each direct set of covariances between data of a given type, $C_{ij}(\mathbf{h})$, for $i = j$, and n_i x n_j submatrices on the off-diagonals for each set of cross-covariances $i \neq j$. The composite LHS matrix is symmetric. For the given data, it may be written as follows:

$$\begin{bmatrix} \left[C_{Z_1Z_1}(\mathbf{u}_\alpha,\mathbf{u}_\beta)\right] & \left[C_{Z_1Z_2}(\mathbf{u}_\alpha,\mathbf{u}_\beta)\right] & \left[C_{Z_1Y}(\mathbf{u}_\alpha,\mathbf{u}_\beta)\right] \\ \left[C_{Z_2Z_1}(\mathbf{u}_\alpha,\mathbf{u}_\beta)\right] & \left[C_{Z_2Z_2}(\mathbf{u}_\alpha,\mathbf{u}_\beta)\right] & \left[C_{Z_2Y}(\mathbf{u}_\alpha,\mathbf{u}_\beta)\right] \\ \left[C_{YZ_1}(\mathbf{u}_\alpha,\mathbf{u}_\beta)\right] & \left[C_{YZ_2}(\mathbf{u}_\alpha,\mathbf{u}_\beta)\right] & \left[C_{YY}(\mathbf{u}_\alpha,\mathbf{u}_\beta)\right] \end{bmatrix} \tag{9.6}$$

where each component of the composite matrix in Equation (9.6) is also a matrix. The term $\left[C_{Z_1Z_1}(\mathbf{u}_\alpha,\mathbf{u}_\beta)\right]$ is an n_{z1} x n_{z1} matrix of data type Z_1 only, so in this case, $\left[C_{Z_1Z_1}(\mathbf{u}_\alpha,\mathbf{u}_\beta)\right]$ is a 3x3 matrix that considers the Z_1 data at locations $\mathbf{u}_1$ to $\mathbf{u}_3$ (see Figure 9.4). Similarly, $\left[C_{Z_1Z_2}(\mathbf{u}_\alpha,\mathbf{u}_\beta)\right]$ is a 3x3 matrix between Z_1 data to Z_2 data at locations $\mathbf{u}_1$ to $\mathbf{u}_3$ using the cross-covariance between Z_1 and Z_2. The term $\left[C_{Z_1Y}(\mathbf{u}_\alpha,\mathbf{u}_\beta)\right]$ is a 3x4 matrix that relates Z_1 data to Y data at locations $\mathbf{u}_1$ to $\mathbf{u}_3$ and $\mathbf{u}_0$. Note that $\left[C_{YZ_1}(\mathbf{u}_\alpha,\mathbf{u}_\beta)\right]$ is the transpose of $\left[C_{Z_1Y}(\mathbf{u}_\alpha,\mathbf{u}_\beta)\right]$, thus it is a 4x3 matrix. The composite LHS matrix (Equation 9.6) for this configuration should be a 10x10 symmetric matrix.

The RHS data-to-estimated location ($\mathbf{u}_0$) vector for estimating the primary attribute Z_1 may be written

$$\begin{bmatrix} C_{Z_1Z_1}(\mathbf{u}_1,\mathbf{u}_0) \\ C_{Z_1Z_1}(\mathbf{u}_2,\mathbf{u}_0) \\ C_{Z_1Z_1}(\mathbf{u}_3,\mathbf{u}_0) \\ C_{Z_1Z_2}(\mathbf{u}_1,\mathbf{u}_0) \\ C_{Z_1Z_2}(\mathbf{u}_2,\mathbf{u}_0) \\ C_{Z_1Z_2}(\mathbf{u}_3,\mathbf{u}_0) \\ C_{Z_1Y}(\mathbf{u}_1,\mathbf{u}_0) \\ C_{Z_1Y}(\mathbf{u}_2,\mathbf{u}_0) \\ C_{Z_1Y}(\mathbf{u}_3,\mathbf{u}_0) \\ C_{Z_1Y}(\mathbf{u}_0,\mathbf{u}_0) \end{bmatrix} \tag{9.7}$$

Similarly, to solve for the cokriging weights for estimaring Z_2 a RHS covariance vector that is the same as that given in Equation (9.7) with the first subscript changed from Z_1 to Z_2 for each element.

The simple cokriging estimate is given by Equation (9.3) and the estimation variance is obtained as (Goovaerts, 1997, p. 267):

$$\sigma^2_{SCK}(\mathbf{u}) = C_{Z_1Z_1}(0) - \sum_{\alpha=1}^{n_{Z1}} \lambda_\alpha C_{Z_1Z_1}(\mathbf{u}_\alpha - \mathbf{u}) - \sum_{\alpha=1}^{n_{Z2}} \lambda_\alpha C_{Z_1Z_2}(\mathbf{u}_\alpha - \mathbf{u}) - \sum_{\alpha=1}^{n_Y} \lambda_\alpha C_{Z_1Y}(\mathbf{u}_\alpha - \mathbf{u}) \quad (9.8)$$

Once the simple cokriging estimates, $z_1^*(\mathbf{u})$, $z_2^*(\mathbf{u})$, and estimation variances, $\sigma_{z1}^{2\ SCK}(\mathbf{u})$, $\sigma_{z2}^{2\ SCK}(\mathbf{u})$ (i.e., the parameters of the local conditional distributions) are obtained, directly simulate 100 realizations each for Z_1 and Z_2 at location $\mathbf{u}_0$ by drawing a random Gaussian deviate and destandardizing:

$$\begin{aligned} z_1^s(\mathbf{u}_0) &= G^{-1}(p) \cdot \sigma_{z1}^{SCK}(\mathbf{u}_0) + z_1^*(\mathbf{u}_0) \\ z_2^s(\mathbf{u}_0) &= G^{-1}(p) \cdot \sigma_{z2}^{SCK}(\mathbf{u}_0) + z_2^*(\mathbf{u}_0) \end{aligned} \quad (9.9)$$

where $G^{-1}(p)$ is the Gaussian inverse corresponding to a randomly drawn percentile, $p \in [0,1]$, $\sigma_{z_1}^{SCK}$ is the conditional standard deviation of Z_1, and z_1^* is the conditional mean of Z_1.

(ii) The full LMC is often difficult to infer reliably, but a correlation matrix is straightforward to obtain for datasets with an adequately large number of samples of each data type (i.e., more than what is given here!). Consider a correlation matrix corresponding to the LMC in (9.5):

	Z_1	Z_2	Y
Z_1	1	-0.5	-0.4
Z_2	-0.5	1	0.6
Y	-0.4	0.6	1

When using the collocated option, the cross correlation between any two variables is obtained via Equation (9.4). As in Part (i), set up the LHS and RHS covariance matrices. First, simulate Z_1 conditioned on all Z_1 data and the collocated Y data. The LHS covariance matrix for this step should consist of 4x4 matrix corresponding to Z_1 data at locations $\mathbf{u}_1$ to $\mathbf{u}_3$ and the Y data at location $\mathbf{u}_0$. The RHS vector is a 4x1 vector relating the three Z_1 data and collocated Y data to the location of interest, $\mathbf{u}_0$. Solve for the kriging weights, calculate the corresponding local mean and variance, and draw simulated values for $Z_1(\mathbf{u}_0)$.

Next, perform cosimulation of Z_2 conditioned on all Z_2 data, the collocated Y data and the collocated Z_1 simulated value previously obtained. Once again, construct the LHS covariance matrix for this simulation; it will be a 5x5 matrix corresponding to Z_2 data at locations $\mathbf{u}_1$ to $\mathbf{u}_3$, Z_1 simulated value at location $\mathbf{u}_0$ and the Y data at location $\mathbf{u}_0$. The RHS vector is a 5x1 vector relating the three Z_2 data and collocated Z_1 and Y data to the location of interest, $\mathbf{u}_0$. Solve for these kriging weights, calculate the corresponding local mean and variance, and draw simulated values for Z_2 ($\mathbf{u}_0$).

Note that each realization of $Z_2(\mathbf{u}_0)$ depends on the preceding simulation for $Z_1(\mathbf{u}_0)$. Once the parameters of the conditional distributions are calculated, it may be easier to draw the simulated values in a spreadsheet application to see the dependency of the simulated values.

Solution

(i) The LHS covariance matrix is

		Z1			Z2			Y			
		1	2	3	1	2	3	1	2	3	0
Z1	1	1.000	0.479	0.376	-0.500	-0.276	-0.226	-0.400	-0.119	-0.075	-0.179
	2	0.479	1.000	0.323	-0.276	-0.500	-0.194	-0.119	-0.400	-0.065	-0.081
	3	0.376	0.323	1.000	-0.226	-0.194	-0.500	-0.075	-0.065	-0.400	-0.129
Z2	1	-0.500	-0.276	-0.226	1.000	0.696	0.602	0.600	0.287	0.226	0.357
	2	-0.276	-0.500	-0.194	0.696	1.000	0.517	0.287	0.600	0.194	0.239
	3	-0.226	-0.194	-0.500	0.602	0.517	1.000	0.226	0.194	0.600	0.299
Y	1	-0.400	-0.119	-0.075	0.600	0.287	0.226	1.000	0.261	0.150	0.419
	2	-0.119	-0.400	-0.065	0.287	0.600	0.194	0.261	1.000	0.129	0.162
	3	-0.075	-0.065	-0.400	0.226	0.194	0.600	0.150	0.129	1.000	0.288
	0	-0.179	-0.081	-0.129	0.357	0.239	0.299	0.419	0.162	0.288	1.000

The RHS covariance vectors for obtaining the weights to estimate Z_1 and Z_2 are given

		Z1	Z2
Z1	1	0.594	-0.326
	2	0.398	-0.238
	3	0.499	-0.284
Z2	1	-0.326	0.769
	2	-0.238	0.633
	3	-0.284	0.710
Y	1	-0.179	0.357
	2	-0.081	0.239
	3	-0.129	0.299
	0	-0.400	0.600

The weights are

		Z1	Z2
Z1	1	0.440	0.004
	2	0.093	0.005
	3	0.310	0.014
Z2	1	-0.004	0.464
	2	-0.005	0.122
	3	-0.012	0.386
Y	1	0.168	-0.183
	2	0.037	-0.041
	3	0.122	-0.141
	0	-0.379	0.417

The estimate (Equation 9.3) and estimation variance (Equation 9.8) giving the mean and variance of the ccdfs at location $\mathbf{u}_0$ are

	$Z_1(\mathbf{u}_0)$	$Z_2(\mathbf{u}_0)$
estimate	-0.652	0.702
est. var	0.438	0.166

(ii) Simulation of Z_1 under a Markov model using collocated cokriging requires conditioning by all Z_1 data and the collocated Y data resulting in a 4x4 LHS covariance matrix and a 4x1 RHS vector:

LHS Covariance

		Z1			Y
	u	1	2	3	0
Z1	1	1.000	0.479	0.376	-0.238
	2	0.479	1.000	0.323	-0.159
	3	0.376	0.323	1.000	-0.199
Y	0	-0.238	-0.159	-0.199	1.000

RHS Covariance

		Z1
	u	0
Z1	1	0.594
	2	0.398
	3	0.499
Y	0	-0.400

Recall that the cross-covariance between Y and Z_1 is given by Equation (9.4). Solving this system gives the following weights:

Weights

	u	Z1
Z1	1	0.395
	2	0.081
	3	0.276
Y	0	-0.238

The corresponding estimate and estimation variance for $Z_1(\mathbf{u}_0)$ are −0.493 and 0.500, respectively. These parameters are used in simulating $Z_1(\mathbf{u}_0)$ via Equation (9.9). Some simulated values for $Z_1(\mathbf{u}_0)$ and $Z_2(\mathbf{u}_0)$ are given below.

Simulation of $Z_2(\mathbf{u}_0)$ via the Markov model now requires all three Z_2 data at locations $\mathbf{u}_1$ to $\mathbf{u}_3$, the collocated Y datum, and the collocated Z_1 realization. To solve for the simple cokriging weights, we do not require the simulated Z_1 value, since the weights are dependent only on the data locations and the correlation between Z_2 and Y. The corresponding LHS and RHS covariance matrices are given below:

LHS Covariance

		Z2			Y	Z1
	u	1	2	3	0	0
Z2	1	1.000	0.696	0.602	0.462	-0.385
	2	0.696	1.000	0.517	0.380	-0.316
	3	0.602	0.517	1.000	0.426	-0.355
Y	0	0.462	0.380	0.426	1.000	-0.400
Z1	0	-0.385	-0.316	-0.355	-0.400	1.000

RHS Covariance

		Z2
	u	0
Z2	1	0.769
	2	0.633
	3	0.710
Y	0	0.600
Z1	0	-0.500

Again, recall that the cross-covariance between Y and Z_2, and between Z_1 and Z_2 takes the same form as Equation (9.4). The corresponding weights to assign the conditioning data are

	u	Z2
Z2	1	0.375
	2	0.096
	3	0.297
Y	0	0.210
Z1	0	-0.136

Calculate directly from these weights the estimation variance of $Z_2(\mathbf{u}_0)$ as 0.246. Given below are five example simulated values for Z_1, the corresponding estimate for Z_2, and five simulated values for Z_2 (see Equation 9.9).

random number	$Z_1^s(u_0)$	$Z_2^*(u_0)$	random number	$Z_2^s(u_0)$
0.2926	-0.8784	0.5909	0.3965	0.4607
0.7429	-0.0944	0.4840	0.9981	1.9210
0.2193	-1.0769	0.6180	0.1287	0.0557
0.7303	-0.1210	0.4877	0.4276	0.3971
0.3129	-0.8793	0.5911	0.2649	0.2791

where the superscript * denotes the estimate and the superscript s denotes the simulated value. Note that the estimation variance is obtained without regard to the simulated value for $Z_1(\mathbf{u}_0)$, which is repeated to yield 100 realizations for $Z_1(\mathbf{u}_0)$ and $Z_2(\mathbf{u}_0)$.

Figure 9.5 shows a comparison of the distribution of simulated values from full and collocated cosimulation. These results will vary slightly from one run to another depending on the random numbers generated. The mean and variance of these distributions should be close to the cokriged results.

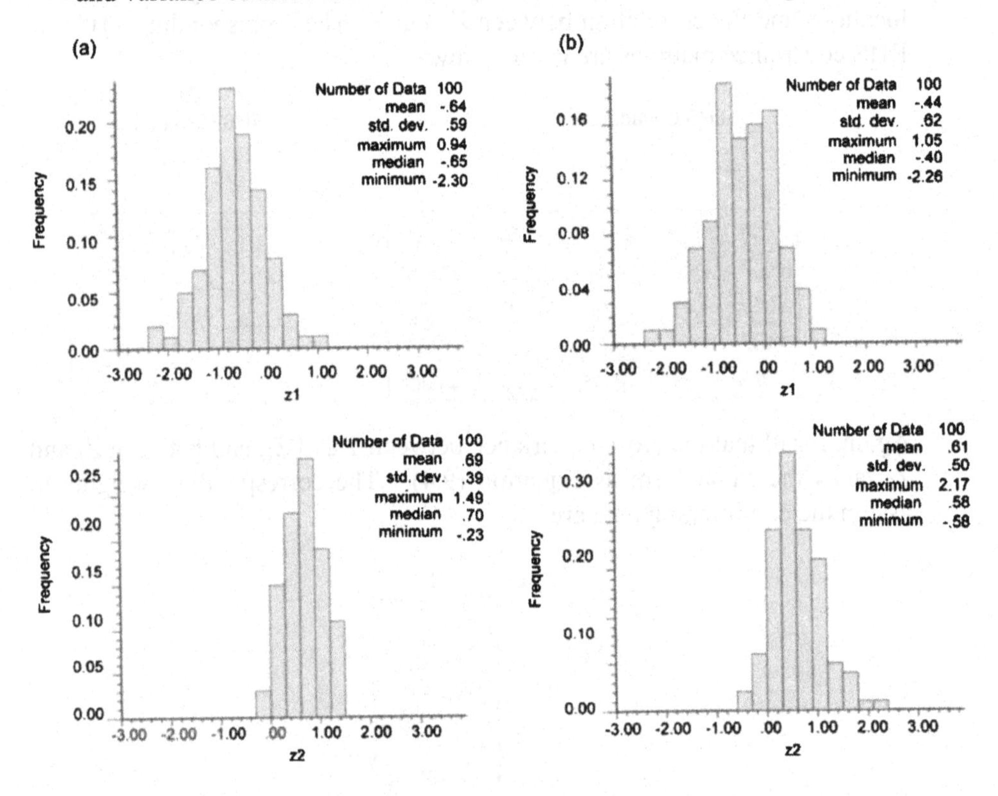

Figure 9.5 Comparison of histograms of simulated values for Z_1 and Z_2 from full cosimulation [column (a)] and collocated cosimulation [column (b)].

Remarks

If all variables existed at the same location, the simulation order may be done on the basis of the variables that are better correlated. This hierarchical scheme is often adopted for collocated cosimulation; that is, an exhaustive grid of one variable is first constructed, and then used as a secondary attribute in collocated cosimulation of the next attribute.

A different form of the Markov model can be adopted in cases where the cross correlogram is likely to mimic the form of the secondary data (Shmaryan and Journel, 1999). This happens most often when the scale of the secondary data is much larger than the scale of the primary data. In this case, $\rho_{12}(\mathbf{h})$ is obtained by a scaling of the correlogram of the secondary data $\rho_{22}(\mathbf{h})$.

Note the variance inflation artifact in the collocated results (see Figure 9.5). This artifact is from retaining only the collocated datum of the secondary variable for the cokriging. This variance inflation should be corrected by applying a variance reduction factor (Deutsch and Journel, 1998; p. 75). In this case, note that the variance of the simulated $Z_2(\mathbf{u}_0)$ from part (ii) is nearly double the estimation variance for $Z_2(\mathbf{u}_0)$ calculated from the full cokriging estimation.

9.3 MULTISCALE COKRIGING

Learning Objectives

Cokriging is a means of estimation simultaneously accounting for data of multiple types. It is also possible to consider data of different *scales* in both kriging and cokriging. This is of interest because there is often secondary information available at different scales and locations. The theoretical advantage of cokriging with secondary data is that the error variance of the cokriging estimator is always equal to or smaller than that of the kriging estimator that considers only the primary variable of interest (Goovaerts, 1997, p.215). There exists a broad spectrum of application of the generalized cokriging formalism. The previous problem showed how multiple secondary data might be combined for simulation. This problem focuses on the integration of data at two different scales in the estimation at an intermediate model grid scale.

Background and Assumptions

The development of the generalized cokriging estimator is worthwhile; although it is but an extension of the kriging formalism (Myers, 1982). Suppose there are P variables, Z_i, $i = 1, \ldots, P$ with mean μ_i defined on volume support V_i centered at location $\mathbf{u}_{\alpha p}$, where $\alpha = 1, \ldots, n_p$ and n_p is the number of available data of type p. The volume scales, V_i, associated with the P variables may be quite different. In fact, the data of one type p may be at different scales. Although the data may be at arbitrarily different volumetric support, the spatial correlation of each variable must be known at an elementary point scale.

Simple or unconstrained kriging is common. It is also common to consider ordinary kriging where the weights are constrained to permit local estimation of the mean (see Problem 6.2). Ordinary cokriging is problematic. Traditionally, ordinary cokriging consisted of enforcing the weights assigned to the primary data to sum to 1.0 and the weights of the secondary variable to sum to 0.0. The constraint on the secondary data weights is restrictive. Any positive weights must be counterbalanced by negative weights; as a result, all secondary data weights end up close to zero. Secondary data have little influence using ordinary cokriging. Simple cokriging is universally recommended.

In simple cokriging, the residuals, Y_p, of the original Z_p variables are used for estimation:

$$Y_p(\mathbf{u}_{\alpha p}) = Z_p(\mathbf{u}_{\alpha p}) - \mu_p(\mathbf{u}_{\alpha p}), \ \forall p, \mathbf{u}_{\alpha p}$$

The volume support of the mean is much larger than the volume support of any data type. The residual data Y_p relate to the specified support V_p. The estimator $Y_i^*(\mathbf{u})$ is a weighted linear combination of the P data types, where i can be any one of the P variables,

$$Y_i^*(\mathbf{u}) = \sum_{p=1}^{P} \sum_{\alpha=1}^{n_p} \lambda_{\alpha p} Y_p(\mathbf{u}_{\alpha p}) \tag{9.10}$$

The corresponding estimation variance is

$$\sigma_E^2 = E\{(Y_i(\mathbf{u}) - Y_i^*(\mathbf{u}))^2\} \tag{9.11}$$

Expanding the RHS of Equation (9.11) and minimizing the error variance, σ_E^2, with respect to the weights, λ, gives the simple cokriging system of equations:

$$\sum_{p=1}^{P} \sum_{\beta=1}^{n_p} \lambda_{\beta p} \bar{C}(V_p(\mathbf{u}_{\alpha p}), V_{p'}(\mathbf{u}_{\beta p'})) = \bar{C}(V_i(\mathbf{u}), V_p(\mathbf{u}_{\alpha p})) \tag{9.12}$$

where $\bar{C}$ indicates average covariance between two volumes of possibly different support. The symmetric covariance matrix in the LHS of Equation (9.12) consists of P x P submatrices of volume-to-volume covariances between different data types,

$$\left[[\bar{\mathbf{C}}(\mathbf{V_p}, \mathbf{V_{p'}})], p, p' = 1, \ldots, P\right] \tag{9.13}$$

where each submatrix consists of n_p x $n_{p'}$ covariances between the p and the p' data.

$$\bar{\mathbf{C}}(\mathbf{V_p},\mathbf{V_{p'}})=\begin{bmatrix}\bar{C}(V_p(\mathbf{u}_{1p}),V_{p'}(\mathbf{u}_{1p'})) & \cdots & \bar{C}(V_p(\mathbf{u}_{1p}),V_{p'}(\mathbf{u}_{n_{p}p'})) \\ \vdots & \ddots & \vdots \\ \bar{C}(V_p(\mathbf{u}_{n_pp}),V_{p'}(\mathbf{u}_{1p'})) & \cdots & \bar{C}(V_p(\mathbf{u}_{n_pp}),V_{p'}(\mathbf{u}_{n_pp'}))\end{bmatrix} \qquad (9.14)$$

The column vectors of weights and RHS covariances in Equation (9.12) consist of $\sum_{p=1}^{P} n_p$ elements:

$$\lambda=\begin{bmatrix}\lambda_{11}\\ \vdots \\ \lambda_{n_1 1} \\ \vdots \\ \vdots \\ \lambda_{1P} \\ \vdots \\ \lambda_{n_P P}\end{bmatrix} \qquad \bar{\mathbf{C}}(\mathbf{V}_i(\mathbf{u}),\mathbf{V}_p(\mathbf{u}_{\alpha p}))=\begin{bmatrix}\bar{C}(V_i(\mathbf{u}),V_1(\mathbf{u}_{11}))\\ \vdots \\ \bar{C}(V_i(\mathbf{u}),V_1(\mathbf{u}_{n_1 1})) \\ \vdots \\ \vdots \\ \bar{C}(V_i(\mathbf{u}),V_P(\mathbf{u}_{1P})) \\ \vdots \\ \bar{C}(V_i(\mathbf{u}),V_P(\mathbf{u}_{n_P P}))\end{bmatrix} \qquad (9.15)$$

Solving for the weights in the cokriging system of equations [Equation (9.12)] results in the following minimum error simple cokriging (SCK) estimate $Y_i^*(\mathbf{u})$ and estimation variance σ_E^2 corresponding to the conditional expectation and variance of the RV $Y_i(\mathbf{u})$ (Goovaerts, 1997, pp. 205–207):

$$Y_1^*(\mathbf{u})=\sum_{\alpha_1=1}^{n_1}\lambda_{\alpha 1}Y_1(\mathbf{u})+\sum_{p=2}^{P}\sum_{\alpha_p=1}^{n_p}\lambda_{\alpha_p}Y_i(\mathbf{u}_{\alpha_p}) \qquad (9.16)$$

$$\sigma^2(\mathbf{u})=C_{11}(0)-\sum_{p=1}^{P}\sum_{\alpha_p=1}^{n_p}\lambda_{\alpha_p}C_{p1}(\mathbf{u}_{\alpha_p}-\mathbf{u}) \qquad (9.17)$$

As with simple kriging, SCK can easily be extended to ordinary cokriging (OCK) by imposing constraints on the weights assigned to the conditioning data to ensure unbiasedness of the OCK estimator; however, as mentioned above ordinary cokriging with the traditional unbiasedness constraints causes the secondary data to have little influence. Moreover, simple cokriging is the theoretically correct approach to estimate the conditional mean and variance in a multivariate Gaussian context (see Problem 7.1).

A variant of OCK is standardized OCK, in which a single constraint is imposed requiring that the sum of all the weights assigned to the primary residual data and all secondary residual data equals 1.0. This approach was presented in Isaaks and Srivastava (1989) and leads to estimates better than conventional OCK with secondary data giving more influence to the estimated value at a location.

Problem

Consider the case where there are porosity measurements from core representing a point-scale primary variable and seismic acoustic impedance data available as secondary information at a block scale of 50 x 50 m (see Figure 9.6):

Primary Core Data

Easting	Northing	Value
22.5	43.5	-0.1877
9.5	24.5	-0.974
24.5	6.5	-0.1876
47.5	20.5	1.215

Secondary Seismic Data

Easting	Northing	Value
25	25	0.2797

The seismic data are correlated to the primary core porosity data with a correlation coefficient of 0.70. Suppose that both variables are multi-Gaussian and share the same point-scale direct variogram:

$$\gamma(\mathbf{h}) = Sph_{a=60}(\mathbf{h})$$

The cross variogram between the two variables is

$$\gamma(\mathbf{h}) = 0.70 Sph_{a=120}(\mathbf{h})$$

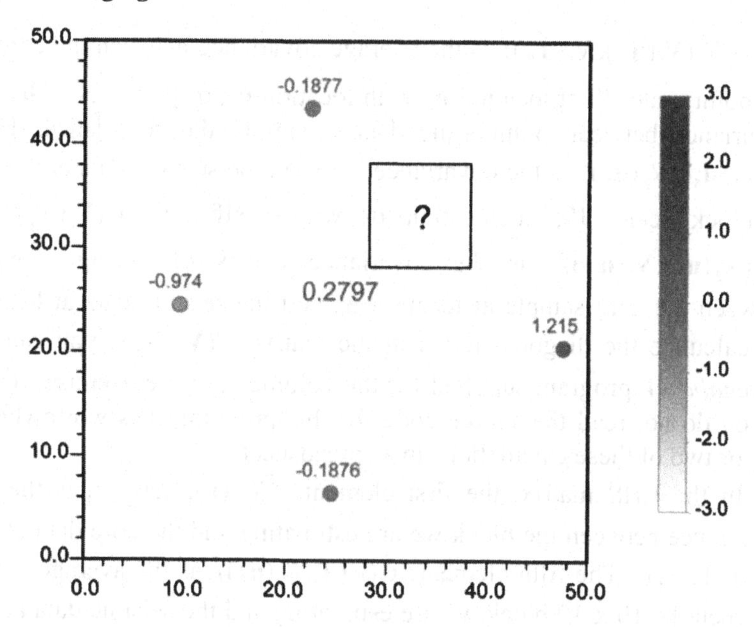

Figure 9.6 Configuration of available data and block of interest.

Using a 3x3 block discretization for all average covariance calculations, calculate and compare the estimated block-averaged porosity and the estimation variance of a 10 x 10 m block centered at location (35, 35) using three approaches:

(i) Write the simple cokriging equations considering these two variables. Use simple cokriging and take advantage of all available information.

(ii) Use ordinary cokriging (OCK) where the sum of the weights of the primary data must sum to 1.0 and the sum of all secondary data must sum to 0.0.

(iii) Use standardized ordinary cokriging where the sum of the weights of the primary and secondary data must sum to 1.0.

Solution Plan

(i) Set up the simple cokriging equations in a spreadsheet as follows:

$$\begin{bmatrix} \bar{C}(V_1(\mathbf{u}_1),V_1(\mathbf{u}_1)) & \bar{C}(V_1(\mathbf{u}_1),V_1(\mathbf{u}_2)) & \bar{C}(V_1(\mathbf{u}_1),V_1(\mathbf{u}_3)) & \bar{C}(V_1(\mathbf{u}_1),V_1(\mathbf{u}_4)) & \bar{C}(V_1(\mathbf{u}_1),V_2(\mathbf{u}_5)) \\ \bar{C}(V_1(\mathbf{u}_2),V_1(\mathbf{u}_1)) & \bar{C}(V_1(\mathbf{u}_2),V_1(\mathbf{u}_2)) & \bar{C}(V_1(\mathbf{u}_2),V_1(\mathbf{u}_3)) & \bar{C}(V_1(\mathbf{u}_2),V_1(\mathbf{u}_4)) & \bar{C}(V_1(\mathbf{u}_2),V_2(\mathbf{u}_5)) \\ \bar{C}(V_1(\mathbf{u}_3),V_1(\mathbf{u}_1)) & \bar{C}(V_1(\mathbf{u}_3),V_1(\mathbf{u}_2)) & \bar{C}(V_1(\mathbf{u}_3),V_1(\mathbf{u}_3)) & \bar{C}(V_1(\mathbf{u}_3),V_1(\mathbf{u}_4)) & \bar{C}(V_1(\mathbf{u}_3),V_2(\mathbf{u}_5)) \\ \bar{C}(V_1(\mathbf{u}_4),V_1(\mathbf{u}_1)) & \bar{C}(V_1(\mathbf{u}_4),V_1(\mathbf{u}_2)) & \bar{C}(V_1(\mathbf{u}_4),V_1(\mathbf{u}_3)) & \bar{C}(V_1(\mathbf{u}_4),V_1(\mathbf{u}_4)) & \bar{C}(V_1(\mathbf{u}_4),V_2(\mathbf{u}_5)) \\ \bar{C}(V_2(\mathbf{u}_5),V_1(\mathbf{u}_1)) & \bar{C}(V_2(\mathbf{u}_5),V_1(\mathbf{u}_2)) & \bar{C}(V_2(\mathbf{u}_5),V_1(\mathbf{u}_3)) & \bar{C}(V_2(\mathbf{u}_5),V_1(\mathbf{u}_4)) & \bar{C}(V_2(\mathbf{u}_5),V_2(\mathbf{u}_5)) \end{bmatrix} \begin{bmatrix} \lambda_1 \\ \lambda_2 \\ \lambda_3 \\ \lambda_4 \\ \lambda_5 \end{bmatrix} = \begin{bmatrix} \bar{C}(V(\mathbf{u}),V_1(\mathbf{u}_1)) \\ \bar{C}(V(\mathbf{u}),V_1(\mathbf{u}_2)) \\ \bar{C}(V(\mathbf{u}),V_1(\mathbf{u}_3)) \\ \bar{C}(V(\mathbf{u}),V_1(\mathbf{u}_4)) \\ \bar{C}(V(\mathbf{u}),V_2(\mathbf{u}_5)) \end{bmatrix}$$

where $\bar{C}(\mathbf{V}_1(\mathbf{u}_\alpha),\mathbf{V}_1(\mathbf{u}_\beta))$ is the average covariance between the core data at the point scale, V_1 at location $\mathbf{u}_\alpha$ with location $\mathbf{u}_\beta$, α, β=1,...,4. The average covariance between point scale data is simply the covariance. The term $\bar{C}(\mathbf{V}_2(\mathbf{u}_5),\mathbf{V}_2(\mathbf{u}_5))$ is the covariance between the seismic data at the 50 x 50 m block scale, V_2, at location $\mathbf{u}_5$ with itself, and $\bar{C}(\mathbf{V}_1(\mathbf{u}_\alpha),\mathbf{V}_2(\mathbf{u}_5))$ $=\bar{C}(\mathbf{V}_2(\mathbf{u}_5),\mathbf{V}_1(\mathbf{u}_\alpha))^T$ is the covariance matrix of average covariances between the core sample at location $\mathbf{u}_\alpha$ and the seismic data at location $\mathbf{u}_5$. To calculate the diagonal terms in the matrix $\bar{C}(\mathbf{V}_2,\mathbf{V}_2)$, you can use the *gammabarvV* program supplied for the volume variance exercise. However, if you do not read the source code for this program, it is worthwhile to do one or two of these calculations in a spreadsheet.

In the RHS matrix, the first element, $\bar{C}(V(\mathbf{u}),V_1(\mathbf{u}_\alpha))$, is the average covariance between the block we are estimating and the core data at location $\mathbf{u}_\alpha$, α=1,...,4. The fifth element, $\bar{C}(V(\mathbf{u}),V_2(\mathbf{u}_5))$, is the average covariance between the 10 x 10 block we are estimating and the seismic data at location $\mathbf{u}_5$.

Note that since we are estimating the value of the primary data at an unsampled location, we use the direct variogram in calculating the first four terms in the RHS covariance matrix. That is, the 10 x 10 m block we are estimating at is considered to be of the same data type as the core data, but having a different volume support. The last term requires the cross variogram for calculating the average covariance between the estimated block and the seismic data.

(ii) Determine and set up the OCK equations given that there are now two constraints: the weights of the primary data must sum to 1.0 and the weight of the secondary data must sum to 0. The resulting OCK system should look like this:

$$\begin{bmatrix} \bar{C}(V_1(\mathbf{u}_1),V_1(\mathbf{u}_1)) & \bar{C}(V_1(\mathbf{u}_1),V_1(\mathbf{u}_2)) & \bar{C}(V_1(\mathbf{u}_1),V_1(\mathbf{u}_3)) & \bar{C}(V_1(\mathbf{u}_1),V_1(\mathbf{u}_4)) & \bar{C}(V_1(\mathbf{u}_1),V_2(\mathbf{u}_5)) & 1 & 0 \\ \bar{C}(V_1(\mathbf{u}_2),V_1(\mathbf{u}_1)) & \bar{C}(V_1(\mathbf{u}_2),V_1(\mathbf{u}_2)) & \bar{C}(V_1(\mathbf{u}_2),V_1(\mathbf{u}_3)) & \bar{C}(V_1(\mathbf{u}_2),V_1(\mathbf{u}_4)) & \bar{C}(V_1(\mathbf{u}_2),V_2(\mathbf{u}_5)) & 1 & 0 \\ \bar{C}(V_1(\mathbf{u}_3),V_1(\mathbf{u}_1)) & \bar{C}(V_1(\mathbf{u}_3),V_1(\mathbf{u}_2)) & \bar{C}(V_1(\mathbf{u}_3),V_1(\mathbf{u}_3)) & \bar{C}(V_1(\mathbf{u}_3),V_1(\mathbf{u}_4)) & \bar{C}(V_1(\mathbf{u}_3),V_2(\mathbf{u}_5)) & 1 & 0 \\ \bar{C}(V_1(\mathbf{u}_4),V_1(\mathbf{u}_1)) & \bar{C}(V_1(\mathbf{u}_4),V_1(\mathbf{u}_2)) & \bar{C}(V_1(\mathbf{u}_4),V_1(\mathbf{u}_3)) & \bar{C}(V_1(\mathbf{u}_4),V_1(\mathbf{u}_4)) & \bar{C}(V_1(\mathbf{u}_4),V_2(\mathbf{u}_5)) & 1 & 0 \\ \bar{C}(V_2(\mathbf{u}_5),V_1(\mathbf{u}_1)) & \bar{C}(V_2(\mathbf{u}_5),V_1(\mathbf{u}_2)) & \bar{C}(V_2(\mathbf{u}_5),V_1(\mathbf{u}_3)) & \bar{C}(V_2(\mathbf{u}_5),V_1(\mathbf{u}_4)) & \bar{C}(V_2(\mathbf{u}_5),V_2(\mathbf{u}_5)) & 0 & 1 \\ 1 & 1 & 1 & 1 & 0 & 0 & 0 \\ 0 & 0 & 0 & 0 & 1 & 0 & 0 \end{bmatrix} \begin{bmatrix} \lambda_1 \\ \lambda_2 \\ \lambda_3 \\ \lambda_4 \\ \lambda_5 \\ \mu \\ \nu \end{bmatrix} = \begin{bmatrix} \bar{C}(V(\mathbf{u}),V_1(\mathbf{u}_1)) \\ \bar{C}(V(\mathbf{u}),V_1(\mathbf{u}_2)) \\ \bar{C}(V(\mathbf{u}),V_1(\mathbf{u}_3)) \\ \bar{C}(V(\mathbf{u}),V_1(\mathbf{u}_4)) \\ \bar{C}(V(\mathbf{u}),V_2(\mathbf{u}_5)) \\ 1.0 \\ 0.0 \end{bmatrix}$$

Where μ and ν are the Lagrange parameters needed to enforce the constraints on the weights (see Goovaerts, 1997, pp. 224–226). Using the weights from this system, calculate the OCK estimate and estimation variance using a slightly modified form of Equations (9.16) and (9.17)

(iii) Set up the standard ordinary cokriging equations given that there is a single constraint that the sum of the weights of the primary data and the weight of the secondary data must sum to 1.0. The resulting standard OCK system should look like:

$$\begin{bmatrix} \bar{C}(V_1(\mathbf{u}_1),V_1(\mathbf{u}_1)) & \bar{C}(V_1(\mathbf{u}_1),V_1(\mathbf{u}_2)) & \bar{C}(V_1(\mathbf{u}_1),V_1(\mathbf{u}_3)) & \bar{C}(V_1(\mathbf{u}_1),V_1(\mathbf{u}_4)) & \bar{C}(V_1(\mathbf{u}_1),V_2(\mathbf{u}_5)) & 1 \\ \bar{C}(V_1(\mathbf{u}_2),V_1(\mathbf{u}_1)) & \bar{C}(V_1(\mathbf{u}_2),V_1(\mathbf{u}_2)) & \bar{C}(V_1(\mathbf{u}_2),V_1(\mathbf{u}_3)) & \bar{C}(V_1(\mathbf{u}_2),V_1(\mathbf{u}_4)) & \bar{C}(V_1(\mathbf{u}_2),V_2(\mathbf{u}_5)) & 1 \\ \bar{C}(V_1(\mathbf{u}_3),V_1(\mathbf{u}_1)) & \bar{C}(V_1(\mathbf{u}_3),V_1(\mathbf{u}_2)) & \bar{C}(V_1(\mathbf{u}_3),V_1(\mathbf{u}_3)) & \bar{C}(V_1(\mathbf{u}_3),V_1(\mathbf{u}_4)) & \bar{C}(V_1(\mathbf{u}_3),V_2(\mathbf{u}_5)) & 1 \\ \bar{C}(V_1(\mathbf{u}_4),V_1(\mathbf{u}_1)) & \bar{C}(V_1(\mathbf{u}_4),V_1(\mathbf{u}_2)) & \bar{C}(V_1(\mathbf{u}_4),V_1(\mathbf{u}_3)) & \bar{C}(V_1(\mathbf{u}_4),V_1(\mathbf{u}_4)) & \bar{C}(V_1(\mathbf{u}_4),V_2(\mathbf{u}_5)) & 1 \\ \bar{C}(V_2(\mathbf{u}_5),V_1(\mathbf{u}_1)) & \bar{C}(V_2(\mathbf{u}_5),V_1(\mathbf{u}_2)) & \bar{C}(V_2(\mathbf{u}_5),V_1(\mathbf{u}_3)) & \bar{C}(V_2(\mathbf{u}_5),V_1(\mathbf{u}_4)) & \bar{C}(V_2(\mathbf{u}_5),V_2(\mathbf{u}_5)) & 1 \\ 1 & 1 & 1 & 1 & 1 & 0 \end{bmatrix} \begin{bmatrix} \lambda_1 \\ \lambda_2 \\ \lambda_3 \\ \lambda_4 \\ \lambda_5 \\ \mu \end{bmatrix} = \begin{bmatrix} \bar{C}(V(\mathbf{u}),V_1(\mathbf{u}_1)) \\ \bar{C}(V(\mathbf{u}),V_1(\mathbf{u}_2)) \\ \bar{C}(V(\mathbf{u}),V_1(\mathbf{u}_3)) \\ \bar{C}(V(\mathbf{u}),V_1(\mathbf{u}_4)) \\ \bar{C}(V(\mathbf{u}),V_2(\mathbf{u}_5)) \\ 1.0 \end{bmatrix}$$

Note that the general estimation variance (Equation 6.5) gives the estimation variance regardless of what type of kriging is used and can be used to compute the estimation variance for all three parts of this problem.

Solution

Computed average covariance values depend on the discretization of the 10x10 block volume and the 50 x 50 seismic block. The LHS average covariance matrix below is based on a 4 x 4 discretization of both block volumes:

LHS

	1	2	3	4	Seismic
1	1.000	0.716	0.552	0.587	0.489
2	0.716	1.000	0.711	0.539	0.503
3	0.552	0.711	1.000	0.669	0.489
4	0.587	0.539	0.669	1.000	0.464
Seismic	0.489	0.503	0.489	0.464	0.694

The shaded covariance terms require the cross-covariance function which is related to the cross-variogram. The RHS covariance terms are not given here and are left to the reader to calculate. It is worthwhile to experiment with the block discretization and take note of the convergence of average variogram values with greater discretization.

The cokriging weights obtained are

	weight
1	0.506
2	0.0296
3	0.0278
4	0.397
seismic	0.0681

The simple cokriging estimate at the 10 x 10 m block centered at (35,35) and the corresponding simple cokriging variance are 0.373 and 0.783, respectively.

As shown in the solution plan, the extension to OCK and standardized OCK is straightforward. The resulting system of equations can be solved using any spreadsheet application. In the case of OCK, where the sum of the weights given to the secondary data must sum to zero, the weight assigned to the seismic data should be zero since only one seismic data is used. In the case of standardized OCK, the sum of the weights assigned to the four point-scale data and the single seismic data must sum to 1.0. The following table compares the results from the three cokriging approaches:

	SCK	OCK	SOCK
estimate	0.057	0.236	0.056
est. var	0.377	0.428	0.377

Remarks

In this problem, both the core and seismic data are multi-Gaussian. In practice, this would require a data transformation. Unfortunately, under the Gaussian transform, the two transformed attributes will not likely average linearly. As such, a cokriging accounting for different scales could result in a bias in the cokriged estimate. For this reason, a multiscale cokriging in the original units of the data would be appropriate if the data average linearly.

CHAPTER 10

Special Topics

This chapter presents some interesting topics that are not as central to the main themes of geostatistics as the preceding chapters; however, geostatistics is a diverse subject area and the aim here is to highlight a few important subjects.

A central theme since the launch of geostatistics has been to calculate best estimates at unsampled locations; however, it is clear that "best" must account for both the uncertainty in the estimate and the consequences of making a mistake. In the presence of complex decisions, it is suggested that the uncertainty in the unsampled value be quantified first, then the best estimate can be established that minimizes the expected loss. The approach has not been widely adopted, but there are important areas of application including grade control in mining.

Another aspect of geostatistics relates to stationarity. Some decision of stationarity is required; otherwise, no estimation or quantification of uncertainty is possible. Trend modeling is considered in the presence of large-scale variations and relatively few data. A great challenge is to model the trend without overfitting the data. Overfitting would lead to poor estimates and too narrow uncertainty. One approach is to build lower dimensional trends, that is a 1D vertical trend and a 2D areal trend, then combine them to a 3D trend. This can produce a reasonable trend model in many cases.

Modeling discrete rock types or facies before considering continuous variables is another approach to address nonstationary features. Stationarity is assumed within the discrete rock types. The challenge is to create realizations of the rock types or facies that reflect realistic geologic complexity. Conventional indicator techniques are limited to variograms/covariances that quantify the relationship between two points at a time. Increasingly, people are interested in using multiple-point statistics that would have the potential to quantify higher order features, such as nonlinear continuity.

Solved Problems in Geostatistics. By O. Leuangthong, K.D. Khan, and C.V. Deutsch

In this chapter, the first problem shows how geostatistical simulation can be used to determine an optimal estimate given a quantified risk position. The second problem deals with schemes to construct a trend as input to geostatistical modeling. The third and final problem is an exercise in multiple point statistics (MPS). Multiple point statistics is a relatively new area in geostatistics, and while it holds much promise for the future of geologic modeling, understanding the information content in MPS is a nontrivial, but key component to MPS application.

10.1 DECISION MAKING IN THE PRESENCE OF UNCERTAINTY

Learning Objectives

Characterizing the uncertainty at an unsampled location has been explored throughout many of the preceding problems. In addition to generating stochastic spatial images, these local probability models can be used as tools for making decisions, for example, the probability of exceeding an important threshold value can be used directly for decision making. In practice, the truth is not known. We have a single estimate and, perhaps, a distribution of uncertainty in what the truth could be. For a given decision, the function describing the consequence or penalty of an incorrect decision may be known and this function can be used to find the '*L*-optimal' estimate, that is, one that minimizes a defined loss function. The objective of this problem is to demonstrate how loss functions can be used in conjunction with geostatistical realizations to obtain best estimates for improved decision making.

Background and Assumptions

Geostatistical models provide kriged estimates and/or a model of uncertainy at unsampled locations. A daunting, but necessary task is then to make decisions based on these inferred value(s). This task is particularly challenging in the presence of uncertainty. The uncertainty may be a parametric Gaussian distribution that can be back transformed, a distribution directly estimated at a series of thresholds, or some number of equally probable realizations. In many applications, it is necessary to have a single value; it is not convenient to directly use uncertainty in final decision making. The challenge is to calculate the optimal value in the presence of uncertainty. Optimal is defined as the value that minimizes future consequences if the decision is made incorrectly.

Loss functions are penalty functions that quantify the consequence of choosing an estimate different from the truth. These functions measure the penalty to both an over and underestimation. If the loss function is symmetric, then a value near the center of the distribution will be retained as the optimal one; however, if the consequences of underestimation and overestimation are quite different, then the optimal value may depart significantly from the center of the distribution. Srivastava (1987) provides an easily understood discussion on loss functions in a geostatistical context.

In the presence of uncertainty, the application of loss functions to determine an optimal estimate involves determining the expected loss of a particular estimate. That is, given a distribution of uncertainty, $F(z)$, and a loss function, $L(z^*)$, the optimal estimate is that which minimizes the expected loss, $E\{L(z^*)\}$. The calculation of $E\{L(z^*)\}$ is based on taking each possible outcome, z, from $F(z)$, and calculating the expected loss for that particular outcome. Thus $E\{L(z^*)\}$ is calculated for all Z in $F(z)$. The outcome that minimizes the expected loss is the L-optimal estimate, where L is a reference to simulated realizations.

Problem

Porosity data from a particular stratigraphic unit are found to fit an exponential distribution with mean m in % porosity:

$$f(z) = \frac{1}{m} e^{-\frac{z}{m}} \text{ with } E\{Z\} = m,\ Var\{Z\} = m^2 \tag{10.1}$$

The multivariate distribution of the stationary porosity random function is assumed multivariate Gaussian after appropriate normal score transform. Covariance analysis on the standard normal scores of porosity suggests an exponential covariance model:

$$Cov\{Y(u), Y(u+\mathbf{h})\} = \rho(\mathbf{h}) = e^{-\frac{|\mathbf{h}|}{a}} \tag{10.2}$$

The covariance is isotropic with integral range $\int_0^\infty \rho(\mathbf{h})d\mathbf{h} = a$ in meters.

Now consider that a particular unsampled location is informed by two nearby samples at locations $\mathbf{u}_1$ and $\mathbf{u}_2$, such that: $|\mathbf{u} - \mathbf{u}_1| < a$ and $|\mathbf{u} - \mathbf{u}_2| < a$. All other samples are too remote to be considered. We wish to characterize the uncertainty about the unsampled value $z(\mathbf{u})$ given the two data values $Z(\mathbf{u}_1) = z_1$ and $Z(\mathbf{u}_2) = z_2$.

The loss function $L(z^*-z)$ for erroneously estimating any value z by z^*, is asymmetric, with a greater penalty for overestimation (Figure 10.1):

$$L(z^* - z) = \begin{cases} (z^* - z)^2 & \text{if } z^* > z \\ |z^* - z| & \text{if } z^* < z \end{cases} \quad (10.3)$$

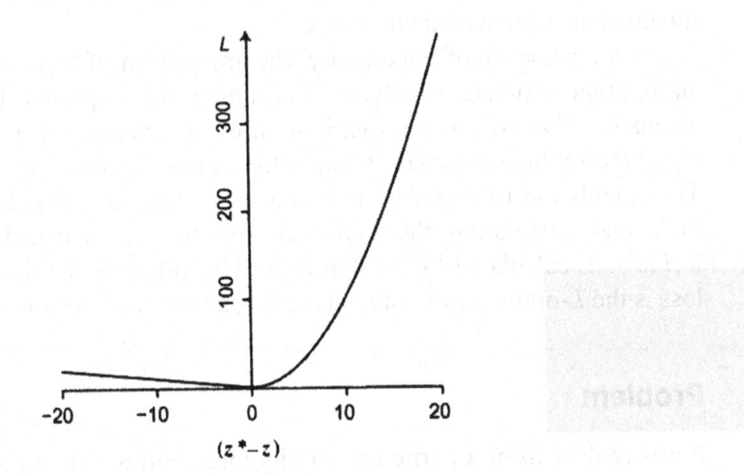

Figure 10.1 Loss function describing the penalty for over and underestimation of *z* by the estimate *z**.

Given: m =10% ; a = 100 m; $|\mathbf{u}-\mathbf{u}_1|=20$ m; $|\mathbf{u}-\mathbf{u}_2|=50$ m; $|\mathbf{u}_1-\mathbf{u}_2|=54$ m; $z(\mathbf{u}_1)$ = 40% (an outlier value); and $z(\mathbf{u}_2)$ = 10%, (i) evaluate the probability that $z(\mathbf{u})$ exceeds the outlier datum value of 40%, and (ii) determine the L-optimal estimate for $z(\mathbf{u})$.

Solution Plan

The local distribution of uncertainty must be obtained. First, this can be accomplished by establishing the link between "z"-space (porosity units) and standard normal y-space via a normal-scores transform of the original, exponentially distributed, porosity data. Second, determine the local (conditional) z-distribution at $\mathbf{u}$ by establishing the parameters of the local normal distribution at $\mathbf{u}$. These parameters are the conditional mean, $m(\mathbf{u})$, and the conditional variance, $\sigma^2(\mathbf{u})$. Finally, perform Monte Carlo simulation to draw normal score values from the nonstandard normal distribution obtained in previous step. Back transform these drawn values to obtain a sample set characterizing the z-conditional distribution of porosity at $\mathbf{u}$. Note that regularly spaced quantiles could also be drawn to back transform values from the conditional Gaussian distribution; however, drawing values by Monte Carlo simulation has the advantage of potentially getting extreme values that may be very important.

The probability of exceeding the porosity threshold value of 40%. This can be determined by a simple count of the simulated values above this threshold, or an examination of the cumulative distribution gives a good approximation.

The *L*-optimal value for the estimate z^* given the loss function provided in Equation (10.3) can also be calculated from the realizations. To do this, consider each possible outcome, z, and set the optimal estimate $z^*=z$. By using this estimate z^*, calculate the loss against each simulated outcome from $F(z)$ determined above. Calculate the expected or average loss for this particular z^*, $E\{L(z^*)\}$. Repeat this for every possible outcome. The *L*-optimal estimate, z^*, is the estimate that minimizes the expected loss, $E\{L(z^*)\}$.

Solution

For the transformation to standard normal space, we have

$$y = G^{-1}(F(z)) \tag{10.4}$$

where $F(z)$ is the cdf expression for the exponential distribution, and $G^{-1}(\cdot)$ is the Gaussian inverse corresponding to the probability $F(z)$. The back-transform to z-space is straightforward given the simplicity of the analytical form for the exponential distribution:

$$z = F^{-1}(G(y)) = -m\ln(1 - G(y)) \tag{10.5}$$

By using Equation (10.4) and the analytical expression for the cdf of the exponential distribution, we transform the two data values at $\mathbf{u}_1$ and $\mathbf{u}_2$ to standard normal scores:

$z(\mathbf{u}_1) = 40$ $\qquad y(\mathbf{u}_1) = G^{-1}(1 - e^{-\frac{40}{10}}) = 2.09$

$z(\mathbf{u}_2) = 10$ $\qquad y(\mathbf{u}_2) = G^{-1}(1 - e^{-\frac{10}{10}}) = 0.34$

The conditional mean, $m(\mathbf{u})$, is given by the simple kriging estimate built from the transformed data, $y(\mathbf{u}_\alpha)$, $\alpha = 1,2$; and the conditional variance, $\sigma^2(\mathbf{u})$, is expanded from the expression for the corresponding error variance:

$$m(\mathbf{u}) = \sum_{\alpha=1}^{2} \lambda_\alpha y(\mathbf{u}_\alpha) \tag{10.6}$$

$$\begin{aligned} \sigma^2(\mathbf{u}) &= C(0) - \sum_{\alpha=1}^{2} \lambda_\alpha C_{\alpha 0} \\ &= 1 - \lambda_1 \cdot \rho_{10} - \lambda_2 \cdot \rho_{20} \end{aligned} \tag{10.7}$$

By calculating correlations via Equation (10.2), we have $\rho_{10} = 0.819$; $\rho_{20} = 0.606$; and $\rho_{12} = 0.583$. Setting up the normal equations to calculate the regression parameters, λ_α, $\alpha = 1, 2$:

$$\begin{aligned} \lambda_1 \rho_{11} + \lambda_2 \rho_{12} &= \rho_{10} \\ \lambda_1 \rho_{12} + \lambda_2 \rho_{22} &= \rho_{20} \end{aligned} \tag{10.8}$$

We obtain $\lambda_1 = 0.705$ and $\lambda_2 = 0.196$. The parameters of the conditional normal distribution at $\mathbf{u}$ are obtained from Equations (10.6) and (10.7), with $m(\mathbf{u}) = 1.540$, and $\sigma^2(\mathbf{u}) = 0.304$.

Given these conditional distribution parameters and that the local distribution is Gaussian, we can draw multiple realizations, say 5000 realizations, of simulated values at location $\mathbf{u}$. The simulated values must then be back transformed to original porosity units using Equation (10.5). These steps can be performed using any spreadsheet application; alternatively, a simple program could be written to perform this task. An example pseudo computer code is shown below to do just this:

Read n*: the desired number of samples*
Read parameters, m *and* $\sigma = \sqrt{\sigma^2}$ *of nonstandard normal distribution*

Loop over n *samples:*
Call random number generator

Call function to compute the inverse of the standard normal cdf = x

y = m + σ *x

$$z = -10.0 * \ln(1.0 - G(y)),$$

where G() *is a function to compute the standard normal cdf given a normal deviate x.*

Write out the n values of z

Note the restandardization step in the pseudo code above, which converts the normal deviate x into a nonstandard normal simulated value, y. In general, this restandardization would also be required in a typical spreadsheet application. The resultant z-distribution is plotted in Figure 10.2. The probability, $\text{Prob}\{z > 40\}$ is calculated from the z-distribution (Figure 10.2) directly (778/5000 = 0.156), or it may be calculated from the transformed distribution as follows:

$$\text{Prob}\{y > y'\} = 1 - G\left\{\frac{y' - m(\mathbf{u})}{\sigma(\mathbf{u})}\right\} = 0.16$$

Using the n drawn samples, we can then establish the loss, in expected value, for a given estimate z^* by

$$E\{L(z^*)\} \cong \frac{1}{n}\sum_{i=1}^{n} L(z^* - z_i) \qquad (10.9)$$

where n is the number of realizations (5000 in these provided solutions).

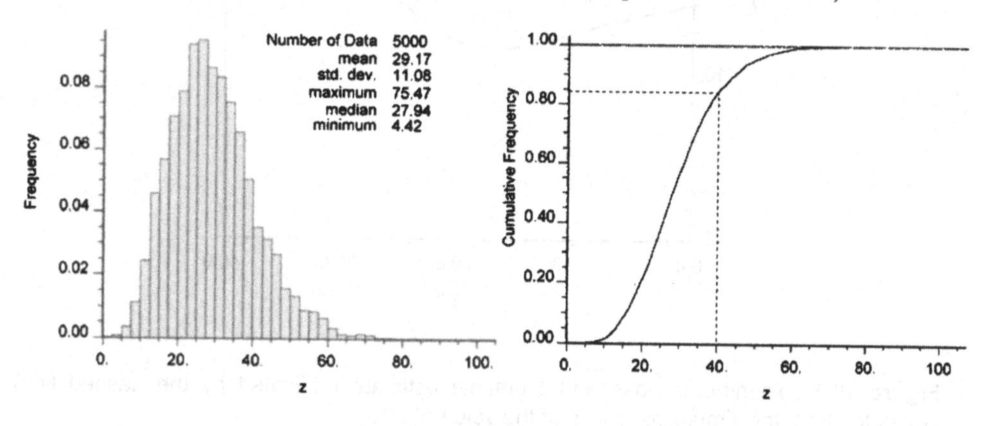

Figure 10.2 Simulated results of the value of $z^*(\mathbf{u})$ from 5000 realizations.

For a series of possible estimates, z^*, the expected loss in Equation (10.9) is calculated. Plotting the expected loss against the possible estimate shows which estimate minimizes the expected loss, thus it is the L-optimal estimate. For this specific case, the L-optimal estimate for porosity at the unsampled location, **u**, is ~16.8% (Figure 10.3).

Remarks

The use of loss functions to calculate estimates that are robust with respect to the anticipated risk/consequences is a standard tool in some decision-making applications. Used in conjunction with geostatistical simulation, it allows us to simultaneously account for the spatial uncertainty of the attribute and determine the best estimate given a risk position. This application is powerful since development decisions are characterized by these two factors: risk and uncertainty.

This formalism has been used for grade control in mining. The decision is the specific disposition of the material mined: processing plant, stockpile, waste dump, and so on. The expected loss is not an actual curve: it will be a single value for every discrete decision that can be taken.

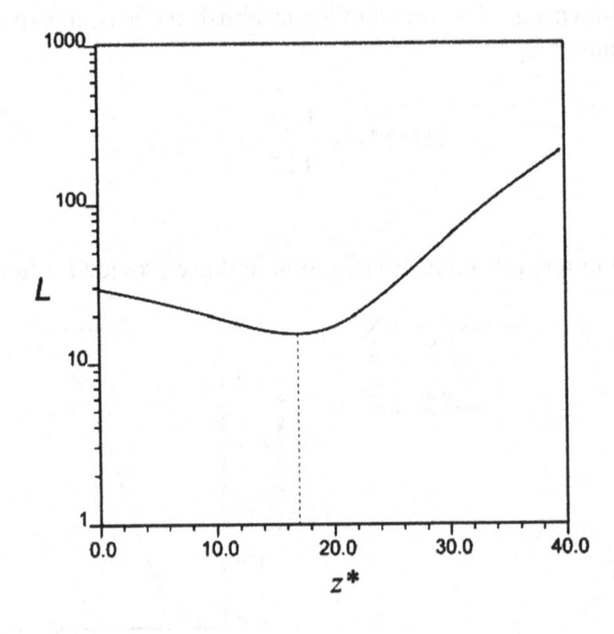

Figure 10.3 Experimental loss and L-optimal estimate (indicated by the dashed line) computed from the simulated results of the value of $z^*(\mathbf{u})$.

10.2 TREND MODEL CONSTRUCTION

Learning Objectives

A decision of stationarity is necessary for all geostatistical tools. The most important aspect of stationarity is first-order stationarity of the mean. Locally dependent second-order moments like the variance, variogram, or covariance are more difficult to assess and are considered of lesser importance. Consider the mean. It could be chosen as a constant known value (SK) or chosen to follow a particular form (OK, UK, or KED) (see Chapter 6). These choices may not be considered appropriate in the presence of a clear trend with sparse data. In this case, we may consider explicitly modeling the locally varying mean. The objective of this problem is to explore the use of probability combination schemes to construct deterministic trend models.

Background and Assumptions

Trends are nonstationary or locally varying statistics of the underlying attribute $Z(\mathbf{u})$. Trends may be inferred from data if there is adequate sampling over the area of interest. If there are many data, then it may not be necessary to use a trend model; local conditioning will enforce the large-scale nonstationary features. If there are too few data, then it may not even be possible to detect a large-scale trend. The use of a trend model is relevant in the presence of relatively few data and a large-scale trend that can be inferred from the data.

A trend should not be fit exactly to the data; this would imply that there are no variations around the fitted trend. Constructing a locally varying trend model that respects large-scale features yet does not overfit the data is a challenge. In 3D spatial analysis, complex trends are rendered tractable by analysis in lower dimensional space (i.e., 1D and 2D). Combining a 1D vertical trend model with a 2D horizontal or areal trend model into a 3D trend model often avoids overfitting. There are many implementation details to consider.

Locally varying mean models can be inferred by a number of methods including moving window statistics or kriging the trend. Under the indicator formalism, the locally varying mean of indicators correspond to local probabilities before updating by local hard or soft data. Computing the average indicator from sample data at a location $\mathbf{u}$ amounts to estimating the conditional probability of that indicator (see Problem 8.3). Consider a single indicator variable $I(\mathbf{u})$. There is an average proportion for $I(\mathbf{u})$ at some 2D location vector (x,y) which may be interpreted as the conditional probability of I given the location event. This is a type of prior probability if we are going to use this conditional information to estimate or simulate a value at this location using this information. Let us denote this indicator probability as P(A|B), where the event A is the local mean or probability of the indicator, and event B is the 2D local probability. Similarly, we may estimate another type of prior indicator probability P(A|C), where event C corresponds to the indicator probability at a 1D location vector defining elevation or depth. These two prior probabilities could be combined to give a probability P(A|B,C) modeling the 3D trend. Bayes' law gives the following expression for the desired conditional probability:

$$P(A \mid B,C) = \frac{P(A,B,C)}{P(B,C)} = \frac{P(A) \cdot P(B \mid A) \cdot P(C \mid A,B)}{P(B,C)} \tag{10.10}$$

The simplest way to combine two probability models is to assume that they are independent, which amounts to simplifying the joint preposteriors to

$$\begin{aligned} P(C \mid A,B) &= P(C \mid A) \\ P(B,C) &= P(B) \cdot P(C) \end{aligned} \tag{10.11}$$

Equation (10.10) can then be written as:

$$P(A \mid B,C) = \frac{P(A) \cdot P(B \mid A) \cdot P(C \mid A)}{P(B) \cdot P(C)} \tag{10.12}$$

Or otherwise expressed as:

$$\frac{P(A \mid B,C)}{P(A)} = \frac{P(A \mid B)}{P(A)} \cdot \frac{P(A \mid C)}{P(A)} \tag{10.13}$$

Equation (10.13) shows clearly how the updating of the global prior probability P(A) is achieved by two independent sources of information. With respect to modeling the trend in the mean, Equation (10.13) can be written as:

$$m(x,y,z) = m_{global} \cdot \left(\frac{m(z)}{m_{global}} \right) \cdot \left(\frac{m(x,y)}{m_{global}} \right) \tag{10.14}$$

An assumption of independence between the two trend components may result in an inconsistent or poorly representative trend model. Locations with a low vertical and areal trend would be estimated very low. Similarly, high areas may be exaggerated. The updated trend value could even exceed 1.0 for indicator variables. An alternative probability combination scheme could be considered.

A probability combination scheme to combine lower order trends to obtain a higher dimensional trend model must satisfy some basic properties (Journel, 2002): (1) the resultant probability must lie within [0,1], that is, $P(A \mid B,C) \in [0,1]$; and (2) the resultant probability and its complement must sum to unity, that is, $P(A \mid (\bullet)) + P(\tilde{A} \mid (\bullet)) = 1$. The first property is a fundamental property of any probability. The second property addresses the issue of closure of probabilities. In the case of full independence, there is no assurance that either of these properties are satisified. In fact, these properties are (1) likely to be violated and (2) unlikely to be verified in the case of full independence. This result is because of the lower dimensional conditional probabilities are evaluated independently.

An alternative to the full independence approach is to assume a model of conditional independence, or some other model of partial data dependency. One practical approach makes use of the permanence of ratios (Journel, 2002), which amounts to assuming that B and C are incrementally conditionally dependent. The approach was proposed as a solution to the closure problem. The model is expressed as:

$$P(A \mid B,C) = \frac{1}{1+x} = \frac{a}{a+bc} = \frac{x-b}{c} = \frac{c-a}{a} \qquad (10.15)$$

Where $\frac{x-b}{c} = \frac{c-a}{a}$ is the permanence of ratios with

$$a = \frac{1-P(A)}{P(A)} \quad b = \frac{1-P(A \mid B)}{P(A \mid B)} \quad c = \frac{1-P(A \mid C)}{P(A \mid C)} \quad x = \frac{1-P(A \mid B,C)}{P(A \mid B,C)}$$

The interpretation of Equation (10.15) is that *the incremental contribution* of information from C to A is independent of B. This result is less restrictive than the assumption of complete independence of B and C.

We introduced this discussion within an indicator framework because the estimated value for an indicator variable obtained by kriging is the proportion of that indicator and is interpreted directly as a probability (see Problem 8.3). The locally varying proportion model constructed using the independence or the permanence of ratios assumptions can be integrated into sequential simulation. The 3D trend model is used directly as a prior mean in indicator kriging. (Deutsch and Journel, 1998, p. 77).

Problem

(i) Construct 1D vertical and 2D horizontal models of the trend in the proportion of each lithofacies, inferred from a dataset of 17 exploratory bore holes, found in the data file *10.2_17wells.dat*. For the construction of the 2D horizontal trend, the binary indicator data are usually averaged along each well and used as conditioning data; these are provided in the data set *10.2_2dwelldata.dat*.

(ii) Combine the two lower order prior models of the nonstationary proportions from Part (i) into a single 3D trend model using (a) the full independence assumption of Equation (10.14), and (b) the permanence of ratios model in Equation (10.15). Build the trend on a 3D grid with extents of 3250 x 3750 x 100 m in *x*, *y*, and *z* directions, respectively. The coordinates of the data file are shifted to a natural origin of (0,0,0) for the Cartesian grid. Check and compare the resultant composite trend model for both cases.

(iii) Simulate via sequential indicator simulation a realization of the indicator variable, conditioned on the data and the full 3D trend model. Compare one indicator realization without the trend model against the one generated with the trend model. The program *sisim_lm* is provided for sequential indicator simulation with locally varying prior proportions.

Solution Plan

(i) Transform the rock type data in the data file *10.2_17wells.dat* into a binary indicator variable of the type (8.1). Compute a vertical moving average of the average indicator value along the Z coordinate. The vertical trend is 1D, so all of the data must be pooled together for a given elevation. For the binary indicator transform (Equation 8.1) of a given facies category, the average indicator identifies the global proportion (i.e., prior probability) of the indicator prevailing at that location or elevation. For a stationary vertical trend, this involves pooling the data together from each elevation, or elevation interval.

In general, trends should be modeled as smoothly varying phenomena since otherwise it is easy to overfit them. Trends can be seen as the deterministic components of the underlying attribute of interest, so any model of the trend is a strong model assumption that should be carefully chosen. Overfitting trends can be avoided by using low-pass filtering (e.g., kriging) or smoothing of the interpolated data. Any interpolation software can be used for these two tasks.

(ii) Using input files consisting of: (1) a column of elevation versus the average indicator, and (2) the gridded 2D trend model, combine the two lower order trend models into a composite 3D trend model. The program *tm_merge.exe* outputs the results from the full independence model and the permanence of ratios model.

(iii) Use any sequential indicator simulation software for Part (iii) and plot the results to visualize the effect of the underlying trend model.

Solution

The well data should be averaged for each elevation interval to yield a single curve describing the varying proportion of each binary indicator as a function of the vertical coordinate (Figure 10.4). Choose a wide enough interval to smooth out high-frequency fluctuations due to natural vertical heterogeneity. The high frequency fluctuations will be added back during simulation.

The curves in Figure 10.4 were obtained by a simple moving window average using a period of 14.5 m. An alternative approach would be a 1D block kriging of the data, or kriging the trend (Goovaerts, 1997, pp. 139–152; Deutsch and Journel, 1998, pp. 66–68). Note that at a given elevation, the sum of the facies proportions equals 1.0, identifying the prior probability model along the vertical coordinate.

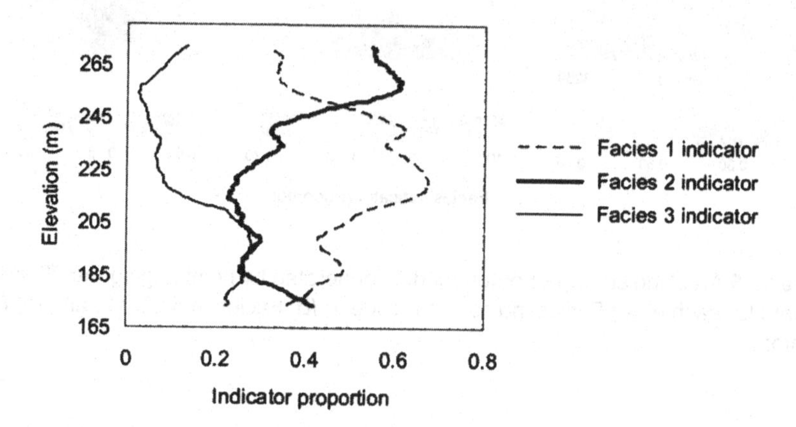

Figure 10.4 Vertical proportion curves for each of the three binary indicator variables generated by a moving window average on the layer-averaged indicator values.

For construction of the 2D areal trend model, the first step is to average each binary facies indicator transform along the well paths to yield the 2D dataset provided in *10.2_2dwelldata.dat*. Ordinary point or block kriging may be used to estimate the average facies proportions at unsampled locations (Figure 10.5). Ordinary kriging is used here to ensure a local estimation of the mean and extrapolate the natural trend across the area.

Combining the lower order prior probability models using a probability combination model based on Equations (10.14 or 10.15) yields a full composite 3D model of the spatial variation in the prior probability of each facies prevailing at a given location (Figure 10.6).

There is little difference upon visual inspection between the composite trend models from the full independence and permanence of ratios assumptions. Histograms of the resultant probabilities do reveal differences worthy of inspection.

Figure 10.7 shows the impact of using a trend model in sequential indicator simulation.

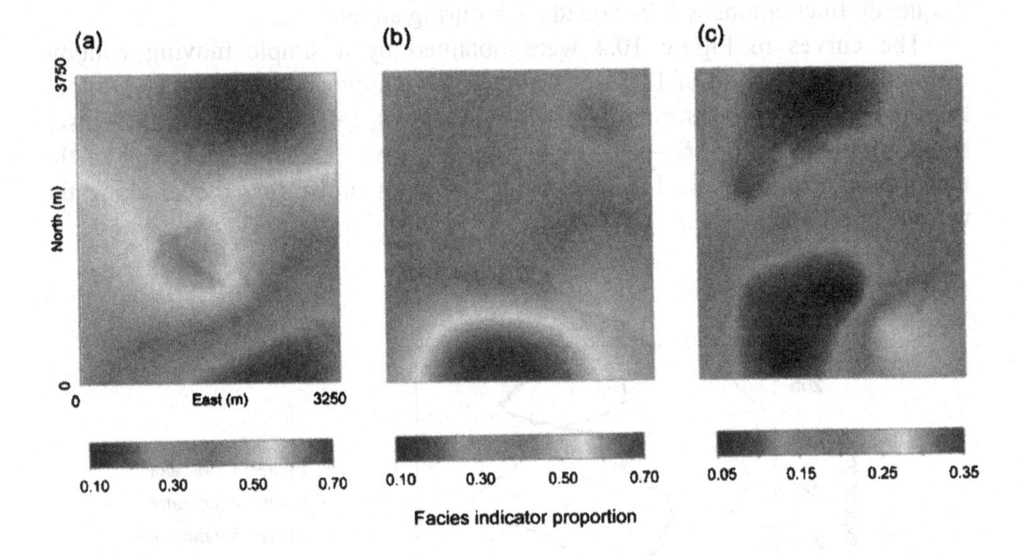

Figure 10.5 Areal indicator proportion models generated by point kriging the 2D averaged well data for each of (a) Facies indicator for code 1; (b) Facies indicator 2; and (c) Facies indicator 3.

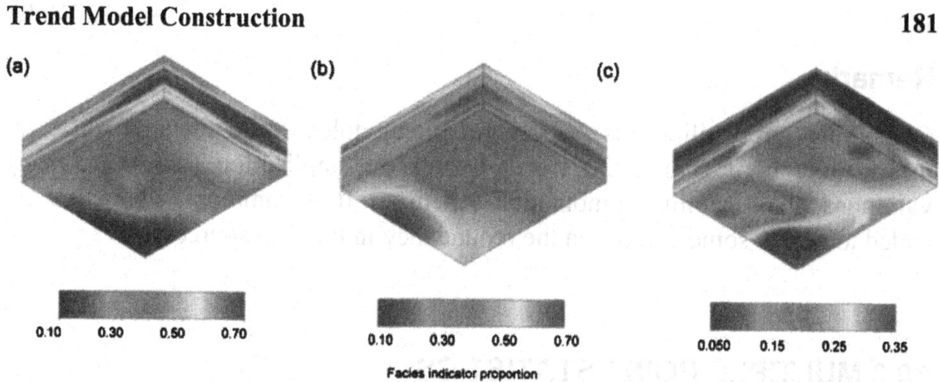

Figure 10.6 Full 3D trend models for (a) Facies code 1; (b) Facies code 2; and (c) Facies code 3 under permanence of ratios assumption.

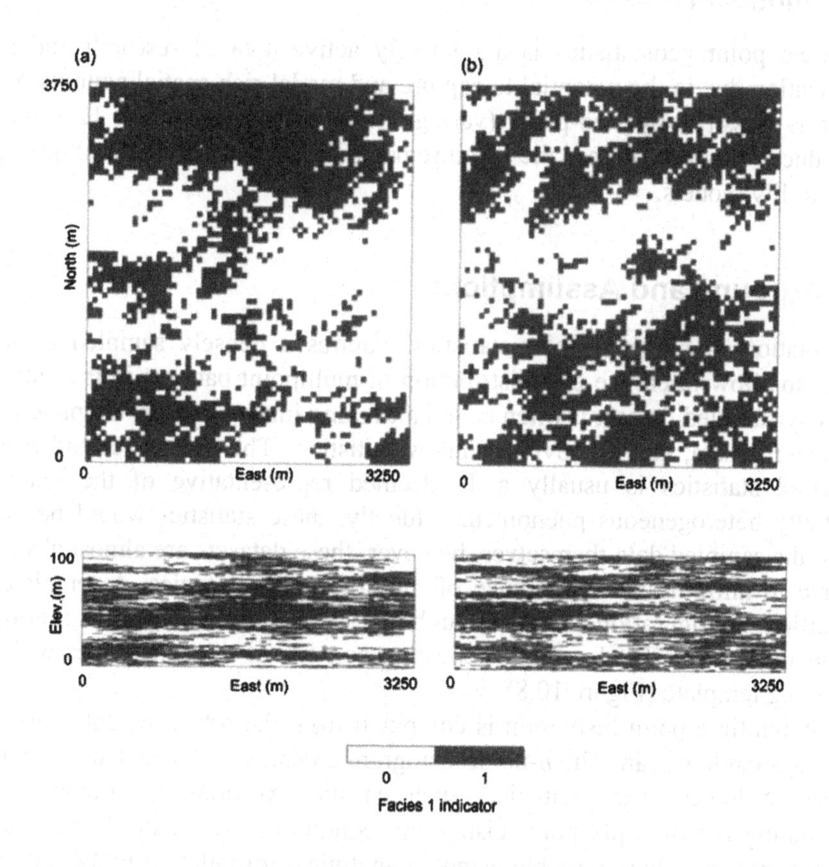

Figure 10.7 (a) Sequential indicator simulation with locally varying prior proportions of the binary indicator corresponding to Facies code 1 shown as x-y areal slice view and x-z vertical cross section view. Note the nonstationary distribution imparted by the trend model by comparing to the same slices of (b) a stationary Sequential indicator simulation.

Remarks

Combining probabilities is a general methodology to integrate different information sources; modeling trends is only one application. There are many variations of probability combination schemes. It is common that factors are added to permit some control on the redundancy in the data sources.

10.3 MULTIPLE POINT STATISTICS

Learning Objectives

Multiple point geostatistics is a relatively active area of research and recent application due to the potential to capture and model rich spatial patterns beyond what is possible by two-point (variogram-based) approaches. This problem introduces and explores the requirements of constructing multiple point probability models.

Background and Assumptions

Calculation of multiple point statistics requires a densely sampled dataset in order to allow inference of a distribution of multipoint patterns in the data. The densely sampled dataset is often called a training image (TI). A template is used to scan the data, like in moving window statistics. The dataset for gathering the required statistics is usually a TI deemed representative of the underlying spatially heterogeneous phenomena. Ideally, these statistics would be derived from the sampled data themselves, however, these datasets are almost always too sparse to allow reliable inference of multiple point statistics. A template of a specified configuration scans through the TI accumulating or counting the frequency of specific, hopefully repeating patterns of discrete values within the scanning template (Figure 10.8).

A multiple point histogram is computed from the reference data based on a moving-window scan. The n-point histogram contains all the statistics, including those of lower order statistics, such as the two-point covariance. After computing the multiple point histogram, conditional simulation using multiple point statistics is then possible using an analytical formulation based on a single normal equation (Guardiano and Srivastava, 1992; Strebelle and Journel, 2001), analogous to kriging, or using iterative techniques (Deutsch, 1992; Caers, 2001).

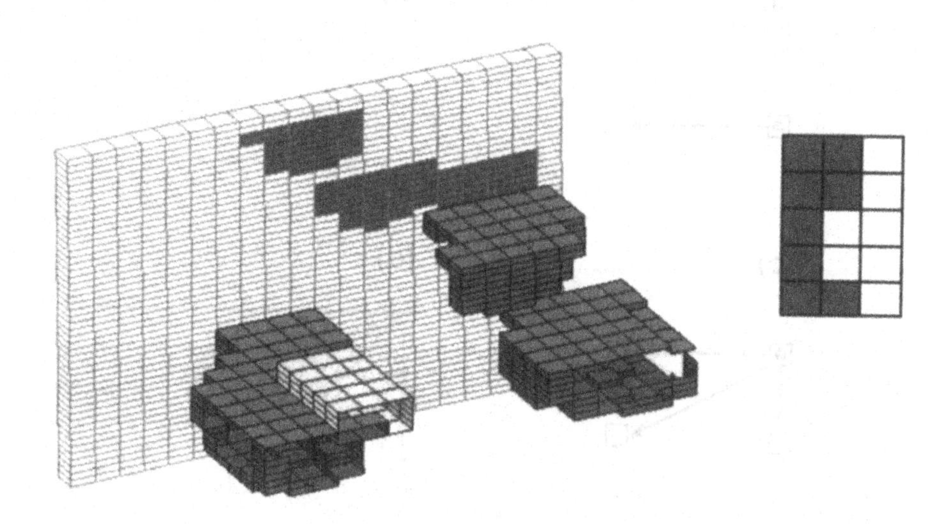

Figure 10.8 The 3D reference data image with geological bodies (gray cells) in background rock (shown as empty space) scanned by 5x3x5 template (white cells). The bottom layer of the multiple point statistic scanned at the current snapshot is shown on the right.

The multiple point histogram is the distribution of the frequency of encountering a specific multiple point data configuration, which is sometimes called a data event, D. The number of unique event classes is K^n, where K is the number of categories (e.g., rock types, property thresholds, pixel shade) in the image and n is the number of elements in the template n-point statistic. These template elements may be defined by a vector of cells, or equivalently, lag vectors from a null vector location, $\mathbf{u}_0$ (Figure 10.9), representing the unknown location for which the conditional probability distribution is required.

For a 2x2 template, there are a total of 4^2=16 unique configurations of a binary or indicator variable (see Figure 10.10). The number of occurrences of each of the unique data configuration as scanned from a prior model constitutes the MP histogram. The unique data events needed to construct the MP histogram can be stored in a list (i.e., a 1D array). Given L categorical variables, $Z_k(\mathbf{u}_l)$ that can take any category in k=1,...K, the data index can be computed as:

$$D_{indx} = 1 + \sum_{l=1}^{L} \left[z_k(\mathbf{u}_l) - 1 \right] \cdot K^{l-1} \tag{10.16}$$

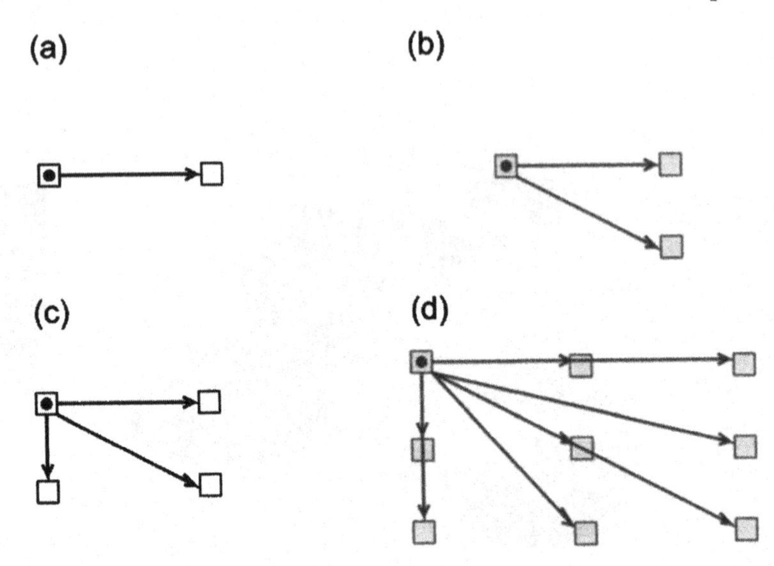

Figure 10.9 Examples of n-point templates defined by an array of ($n-1$) vectors for (a) two; (b);three; (c) four; and (d) nine-point statistics. The null vector $\mathbf{u}_0$ is shown by the black dot.

Computing the conditional probability distribution at a location requires inference of the marginal and joint n-point probabilities. As an example, consider a case with three categorical variables and a four-point template shown in Figure 10.11. The marginal probabilities are obtained as the MP histogram of the data events considering the three locations ($\mathbf{u}_1$, $\mathbf{u}_2$, and $\mathbf{u}_3$). Consider the event that $\mathbf{u}_1$=1, $\mathbf{u}_2$=2, and $\mathbf{u}_3$=2. The conditional probability distribution is evaluated via Bayes' law as

$$\begin{aligned}
f(\mathbf{u}_0 = 1 \mid \mathbf{u}_1 = 1, \mathbf{u}_2 = 2, \mathbf{u}_3 = 2) &= \frac{f(\mathbf{u}_0 = 1, \mathbf{u}_1 = 1, \mathbf{u}_2 = 2, \mathbf{u}_3 = 2)}{f(\mathbf{u}_1 = 1, \mathbf{u}_2 = 2, \mathbf{u}_3 = 2)} \\
f(\mathbf{u}_0 = 2 \mid \mathbf{u}_1 = 1, \mathbf{u}_2 = 2, \mathbf{u}_3 = 2) &= \frac{f(\mathbf{u}_0 = 2, \mathbf{u}_1 = 1, \mathbf{u}_2 = 2, \mathbf{u}_3 = 2)}{f(\mathbf{u}_1 = 1, \mathbf{u}_2 = 2, \mathbf{u}_3 = 2)} \\
f(\mathbf{u}_0 = 3 \mid \mathbf{u}_1 = 1, \mathbf{u}_2 = 2, \mathbf{u}_3 = 2) &= \frac{f(\mathbf{u}_0 = 3, \mathbf{u}_1 = 1, \mathbf{u}_2 = 2, \mathbf{u}_3 = 2)}{f(\mathbf{u}_1 = 1, \mathbf{u}_2 = 2, \mathbf{u}_3 = 2)}
\end{aligned} \tag{10.17}$$

Computing these probabilities in a binary indicator context leads to the use of kriging to derive the conditional distributions (Strebelle, 2002).

All of the marginal and joint probabilities can be scanned from the reference image and stored as the MP histogram. The problem with this direct inference is of course that the number of unique data events to store becomes impractical very quickly since the dimension is K^n.

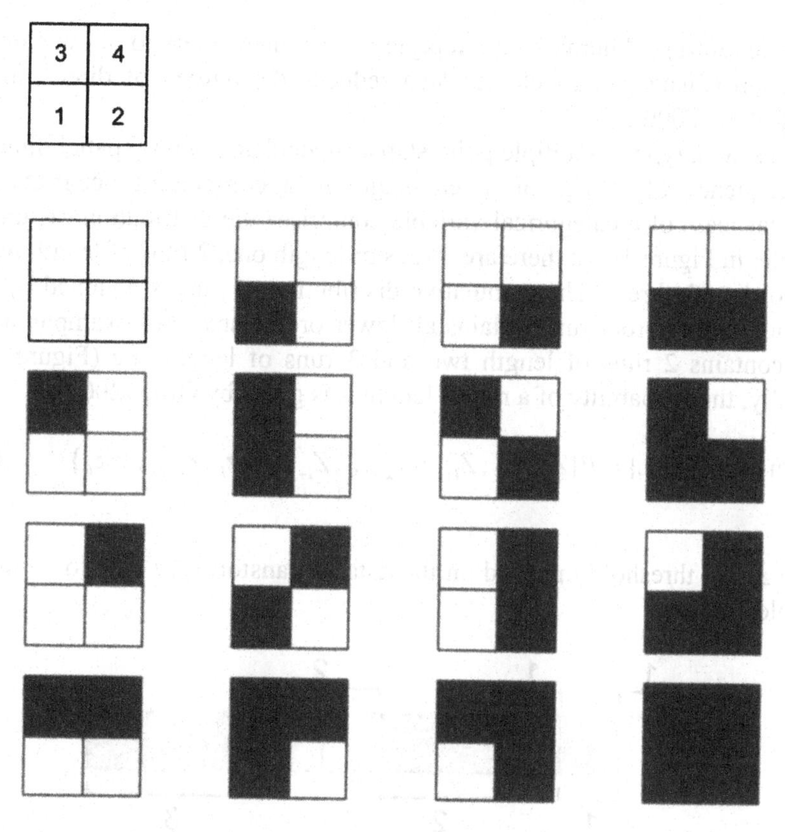

Figure 10.10 The $K^n = 16$ unique data events possible in the case of a binary categorical variable and the four-point template shown here. The index number increases row-wise from 1–16 from the top left.

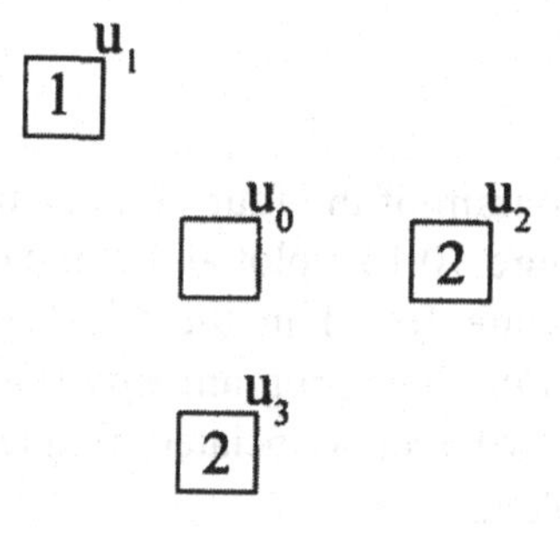

Figure 10.11 Four-point template used to compute the conditional probability distribution at location $\mathbf{u}_0$ based on the data event $\mathbf{u}_1 = 1, \mathbf{u}_2 = 2, \mathbf{u}_3 = 2$.

Nevertheless, it is possible to work with multiple point statistics by restricting K to a small number (less than 4) of sparse data templates, and employing efficient search tree approaches to reduce the computational burden

(Strebelle, 2001). Alternatively, applying directional filters to allow simulation from approximate pattern classes also reduces the burden of direct inference (Zhang et al., 2006).

Runs are a type of multiple point statistic calculated in one spatial dimension. The frequency of strings of given length in a consecutive occurrence of a particular class of a categorical variable comprises the distribution of runs. For example, in Figure 10.12 there are 3 runs of length one, 2 runs of length two, and 1 run of length three. The cumulative distribution of runs is obtained by noting that each higher order run contains all lower order runs. For example, a run of three contains 2 runs of length two and 3 runs of length one (Figure 10.12). Formally, the probability of a run of length L is given by Ortiz (2003):

$$P\{\text{Run of length L}\} = P\{Z_i > z_k, Z_{i+1} \leq z_k, \ldots, Z_{i+L+1} \leq z_k, Z_{i+L+2} > z_k\} \qquad (10.18)$$

where z_k is a threshold imposed on the data to transform the data to an indicator variable.

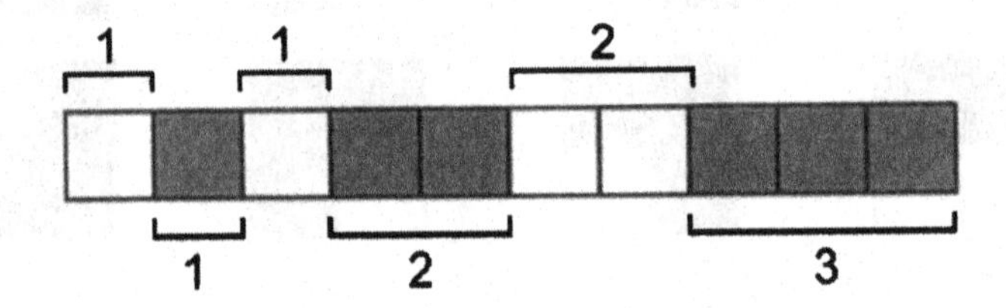

Figure 10.12 Runs of length one, two, and three along a string of model cells populated by two categories.

Problem

(i) Compute the MP histogram of the four-point statistic shown below from two reference images (Figure 10.13); plot and compare the MP histograms. The reference image files are found in the files *10.3_Part1_Image_a.dat* and *10.3_Part1_Image_b.dat*. The program *mphist.exe*, which writes out the 1D array [Equation (10.16)] and associated frequency of each data event is included for this problem.

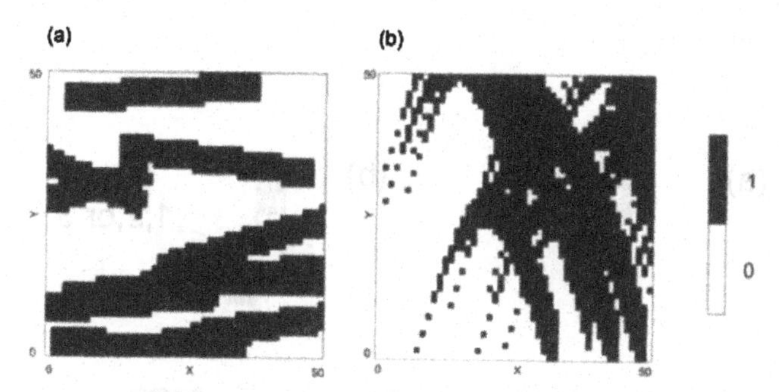

Figure 10.13 Pixelplots of the training images found in the files (a) *10.3_Part1_Image_a.dat* and (b) *10.3_Part1_Image_b.dat*

(ii) Compute the distribution of runs for the same two reference images (Figure 10.13) in each of the coordinate directions and compare the cumulative runs distributions. The program *runsdist.exe* computes the runs in the two directions parallel to the X and Y coordinate axes and writes out the distribution of runs.

(iii) Construct the multiple point conditional probability distribution at an unsampled location, $\mathbf{u}_0$, given the conditioning data event (Figure 10.14a), and training image in the file *10.3_Part3_image.dat* (Figure 10.14). Simulate multiple realizations from this ccdf. Consider a single simulated value as a fourth conditioning datum and compute the conditional probability distribution at the new unsampled location [Figure 10.14(b)] to explore the sequential simulation paradigm in a multiple point context.

Computing the probabilities required in Part (iii) requires assembling the joint and marginal probabilities similar to that shown in Equation (10.17). Write your own one-off computer code to get these probabilities. Alternatively, the program *mppdf.exe* is supplied for computing the statistics based on a four or five point 2D template. Note that the lag vectors of the data templates [Figure 10.14(a,b)] referenced from the unsampled location $\mathbf{u}_0$ are as follows:

(a)

	h(Δx)	h(Δy)	cat
u_1	−1	0	1
u_2	0	2	2
u_3	1	−1	3

(b)

	h(Δx)	h(Δy)	cat
u_1	−1	−1	1
u_2	0	1	2
u_3	1	−2	3
u_4	0	−1	(simulated)

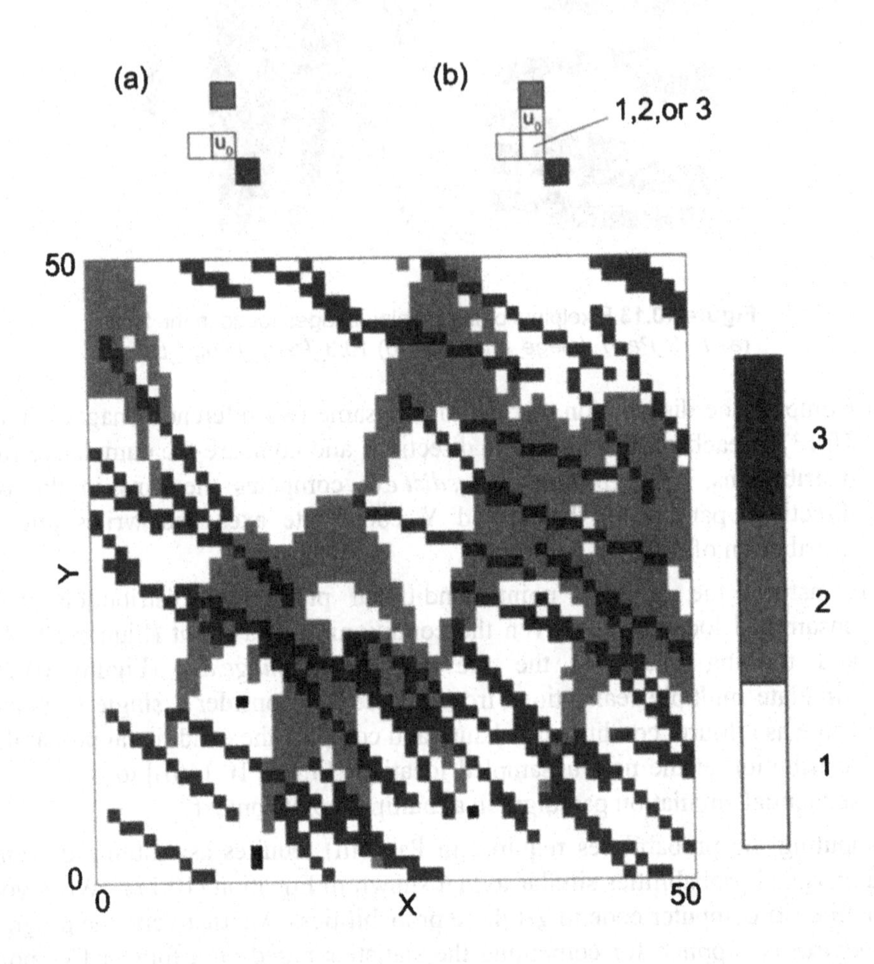

Figure 10.14 Reference image in file *10.3_Part3_image.dat* with (a) four-point template based on a specific data event consisting of three conditioning data and an unsampled location $\mathbf{u}_0$, and (b) a five-point template based on four conditioning data after obtaining a simulated value of category 1,2, or 3 from the previous step (a).

Solution Plan

(i) To compute the MP histogram of a statistic, simply count the frequency of each individual class given by the data index of Equation (10.16).

(ii) Computing the distribution of runs in 1D along a given orientation is a similar bookkeeping computation where all that is stored is the frequency of strings of given length ($1,\ldots,R_{max}$) of any category (threshold) for a categorical (continuous) variable. The distribution can be plotted simply as a histogram, or perhaps more informative is a plot of the cumulative distribution (cdf) of runs.

(iii) Write a short program to compute the probabilities in Equation (10.17).

Solution

(i) The multiple point histograms for the training images of Figure 10.13 are plotted in comparison with Figure 10.15. The multiple point moments corresponding to data indices 1 and 16 (Figure 10.10) dominate the distribution simply because the statistic we are calculating can only provide information on a short length scale due to its small size. Discounting these two classes, the two images are quite different at short scale. The second image [Figure 10.13(b)] is richer in pattern complexity as indicated by the histogram. For example, discounting data indices 1 and 16, note that image (a) is almost entirely comprised of indices 4 and 13 with zero frequency of indices 7 and 10 (Figure 10.10). Image (b) has no zero classes, that is, all of the index patterns are informed.

To do MP geostatistics with the aim of reproducing the reference image, one would have to consider a larger template than the four-point statistic used here for illustrative purposes, but plotting the MP histogram would become impossible to interpret graphically due to the large number of classes.

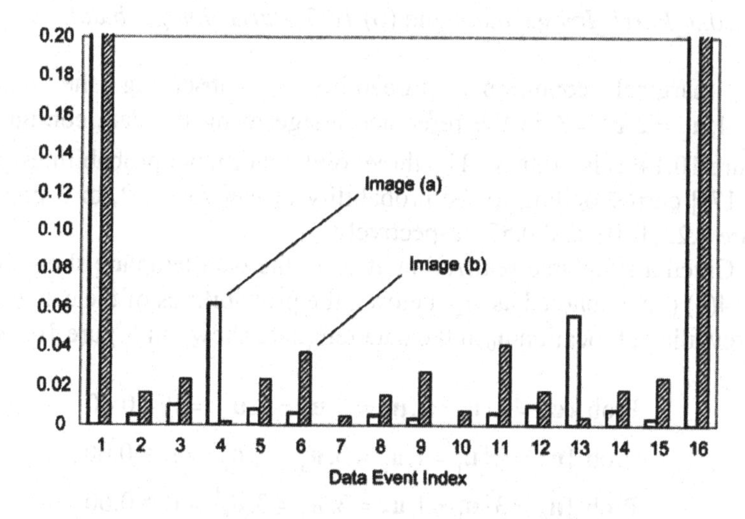

Figure 10.15 Multiple point histogram for the two reference images in the files *10.3_Part1_Image_a.dat* and *10.3_Part1_Image_b.dat*.

(ii) Since the runs are a type of 1D connectivity function, the distribution of runs in Figure 10.16 show a pronounced anisotropy for the strongly directional image in Figure 10.13(a), but greater similarity between the two coordinate directions for image (b). The cumulative distribution of runs for reference image (a) is flat in the *X*-direction and steep in the *Y*-direction with a crossover point between run length 8 and 7. The interpretation is that long run lengths are characteristic of coordinate direction *X* and short runs are dominant in the *Y*-direction, with runs of length 8 having approximately equal probability in either direction.

On the other hand, the distribution of runs is only weakly dependent on direction for image (b).

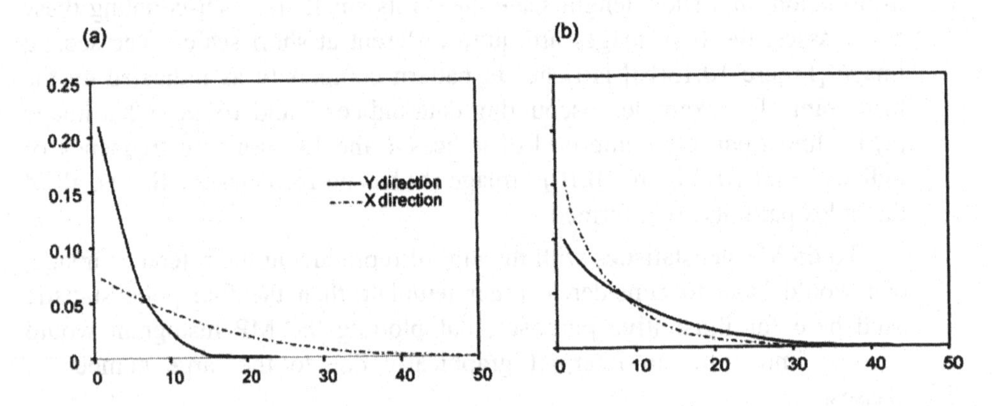

Figure 10.16 Cumulative distribution of runs for the two reference images in the files (a) *10.3_Part1_Image_a.dat* and (b) *10.3_Part1_Image_b.dat.*

(iii) The marginal conditional probability of observing the data event $\mathbf{u}_1 = 1, \mathbf{u}_2 = 2, \mathbf{u}_3 = 3$ in the reference image using the data configuration in Figure 10.14(a) is ~0.093. The three joint conditional probabilities [Equation (10.17)] corresponding to the probability of category 1,2, or 3 prevailing at $\mathbf{u}_0$ are 0.23, 0.19, and 0.57, respectively.

Given a simulated realization at $\mathbf{u}_0$ in the data template shown in Figure 10.14(a) (now denoted as $\mathbf{u}_4^*$ below), the probabilities of the category 1,2, or 3 prevailing at location $\mathbf{u}_0$ in the data template shown in Figure 10.14(b) are

$$\text{Prob } \{\mathbf{u}_0 = 1 \mid \mathbf{u}_1 = 1, \mathbf{u}_2 = 3, \mathbf{u}_3 = 3, \mathbf{u}_4^* = 1\} \cong 0.40$$

$$\text{Prob } \{\mathbf{u}_0 = 2 \mid \mathbf{u}_1 = 1, \mathbf{u}_2 = 3, \mathbf{u}_3 = 3, \mathbf{u}_4^* = 1\} \cong 0.00$$

$$\text{Prob } \{\mathbf{u}_0 = 3 \mid \mathbf{u}_1 = 1, \mathbf{u}_2 = 3, \mathbf{u}_3 = 3, \mathbf{u}_4^* = 1\} \cong 0.60$$

$$\text{Prob }\{\mathbf{u}_0 = 1 \mid \mathbf{u}_1 = 1, \mathbf{u}_2 = 3, \mathbf{u}_3 = 3, \mathbf{u}_4^* = 2\} \cong 0.0$$
$$\text{Prob }\{\mathbf{u}_0 = 2 \mid \mathbf{u}_1 = 1, \mathbf{u}_2 = 3, \mathbf{u}_3 = 3, \mathbf{u}_4^* = 2\} \cong 0.75$$
$$\text{Prob }\{\mathbf{u}_0 = 3 \mid \mathbf{u}_1 = 1, \mathbf{u}_2 = 3, \mathbf{u}_3 = 3, \mathbf{u}_4^* = 2\} \cong 0.25$$

$$\text{Prob }\{\mathbf{u}_0 = 1 \mid \mathbf{u}_1 = 1, \mathbf{u}_2 = 3, \mathbf{u}_3 = 3, \mathbf{u}_4^* = 2\} \cong 0.33$$
$$\text{Prob }\{\mathbf{u}_0 = 2 \mid \mathbf{u}_1 = 1, \mathbf{u}_2 = 3, \mathbf{u}_3 = 3, \mathbf{u}_4^* = 2\} \cong 0.17$$
$$\text{Prob }\{\mathbf{u}_0 = 3 \mid \mathbf{u}_1 = 1, \mathbf{u}_2 = 3, \mathbf{u}_3 = 3, \mathbf{u}_4^* = 2\} \cong 0.50$$

Remarks

The data templates used to compute MP statistics in this problem were very small. Geostatistical modeling with MP statistics requires considerably higher order statistics than those obtained over four to five points. There are many problems associated with reliable inference of high-order statistics. Obtaining or generating the training image is a key challenge. The reference image must be an exhaustive (or very dense) set of (usually synthetic) data containing all of the spatial features that are present, or deemed relevant in the underlying spatial phenomenon. The combinatorial of data configurations (data events) quickly becomes intractable so it is necessary to limit the number of data values (i.e., the K categories or thresholds) in order to use a large enough n-point statistic.

CHAPTER 11

Closing Remarks

The preceding nine chapters present 27 problems related to the fundamentals of geostatistics. Errors and inconsistencies in the problems and solutions will be identified in the future. Corrections and clarification will be documented at www.solvedproblems.com. All required data files and programs will also be available at this website. Our intention is to spend reasonable efforts expanding on the problems started here. We have no doubt that solving relevant problems is the best way to learn a discipline, especially one as diverse as geostatistics.

The aim of many geostatistical studies is uncertainty: local uncertainty in what might be encountered at an unsampled location or global uncertainty in some resource. **Chapter 2** illustrated some basic probability concepts. Virtually all techniques for uncertainty assessment require a prior distribution of uncertainty (cdf or pdf for continuous variables or proportions for categorical variables. **Chapter 3** introduced some techniques to establish representative statistics. Monte Carlo simulation is an approach widely used in many aspects of science and engineering to transfer uncertainty in input variables (point scale rock properties) to output variables (larger scale resources). **Chapter 4** demonstrated the fundamental tool of Monte Carlo Simulation (MCS). A central feature of all geospatial variables is spatial correlation. Geology is not random. The specific pattern of correlation at a particular site must be quantified. **Chapter 5** presented the variogram for the purpose of quantifying spatial variability. Historically, it was of great interest to calculate the best estimate at unsampled locations. *Best* is defined as an estimate that minimizes the expected squared difference between estimate and the unknown truth. **Chapter 6** developed kriging as a technique to calculate best estimates.

Drawing realizations of spatially distributed variables required a multivariate (spatial) probability distribution of the variable. There are only two practical

Solved Problems in Geostatistics. By O. Leuangthong, K.D. Khan, and C.V. Deutsch

approaches: (1) build realizations with some algorithm and observe the distribution, or (2) parameterize the distribution with a multivariate Gaussian distribution. **Chapter 7** focused on Gaussian simulation and its unquestioned importance in modeling uncertainty in continuous variables. Categorical variables are often tackled with indicator techniques. Sometimes continuous variables could be addressed with the indicator formalism. Although the indicator formalism is not used extensively for continuous variables, it is interesting to see an alternative to the Gaussian distribution. **Chapter 8** was devoted to problems of indicator geostatistics. It is rare to be faced with a geostatistical problem with only one variable. There are often multiple variables of interest; some are value adding and some are contaminants. **Chapter 9** presented some problems related to modeling more than one variable simultaneously. The rich variety of geostatistics was not fully covered in Chapters 2–9; **Chapter 10** addressed some special topics.

The topics that have not been covered exceed those that have been covered in the preceding chapters. The rich complexity of geological phenomena, the variety of tools available to address these problems, and the different implementation choices of these tools is large. Some of the important outstanding subjects are mentioned below.

Geostatistics is called on to model heterogeneity and quantify uncertainty; however, there are a number of challenges at the start of a project including the definition of the problem, the scope of the study and the workflow required. Rarely is there a simple recipe that permits reasonable solution to a practical problem.

Geostatistical tools work with measured observations of some variable of interest called *data*. There are inevitable problems with data. The measurements may be inconsistent, in error, systematically biased or problematic in some other way. There are a number of important considerations before the problems of declustering and debiasing are encountered. It is poor practice to take the data at face value and start constructing geostatistical models. Good and best practice would involve a thorough checking and quality control of each available data type. Establishing the data that are without significant errors or biases is a challenge. Deciding how to deal with data of different vintage and scale is nontrivial. These practical issues are difficult to address with neat self-contained problems.

Another practical aspect of geostatistics is to establish a geometric framework within which to perform statistical inference. The large scale geological structure must be visualized and considered in a coordinate system that facilitates the integration of geological data. Often, stratigraphic coordinates or unfolding is considered. The grid may be adapted to faults and other structural complexities. There may be uncertainty in these structural features and controls. The complexities of these structural problems make it difficult to specify neat self-contained problems.

Choosing stationary populations within which to pool data and perform common statistical analysis may be a challenge. Most geological sites subjected to geostatistical analysis are anomalies. There are zones of higher or lower values based on rock types, depositional system or some other features. Choosing to pool disparate data together would lead to poor predictions. Choosing to subdivide the site into too many different populations would lead to difficult spatial inference and poor predictions. A challenge for all geostatistical modeling is an appropriate choice of (1) how to group the data, and (2) location dependence of spatial statistics. This two-part choice constitutes a decision of stationarity.

Separating data by rock type, soil type, facies, lithology, mineralogy, or some other property greatly helps the decision of stationarity. The regionalized variables of interest are often very different within these rock types. The spatial arrangement of rock types is very dependent on the particular geological site being considered. There are many different rock type modeling procedures. The indicator technique presented here is one of the simplest. Truncated pluriGaussian, process mimicking, and object-based modeling are applicable to a variety of settings. A variety of theoretical and practical issues surround these techniques that make it difficult to set neat self-contained problems.

Geostatistical models are constructed to assist in decision making. Decisions require many variables. As mentioned, some variables are value-adding, some are deleterious, and some are rate constants. The complexity of the multivariate distribution between these variables is not fully reflected by the simple techniques revealed in this book. There are many other techniques for multivariate geostatistical problems. Additional problems are warranted to understand these other techniques.

Issues of scale are ubiquitous in geostatistics. Data are often measured at a small scale (10^{-5} m^3). Some geophysical data is at a large scale (10^3 m^3). There may be multiple data types with drastically different scales. Models are constructed at an intermediate scale and, perhaps, used at another scale. Geological variability appears different at different scales. The important scale of variability is very problem specific. This fascinating aspect of geostatistical modeling warrants more problems in this text; however, it is difficult to define neat self-contained problems.

Geostatistical modeling is often undertaken hierarchically. Large-scale structures are modeled first, followed by rock types, followed by continuous variables. The specific work flow is adapted to each problem. Geostatisticians learn the practical work flow for a particular problem. It would depend on the goals of the study, the available data, the available professional and computer resources, and the time available for the study. There is no universal best technique. Our ability to choose the best approach for a particular problem is directly proportion to the number of problems we have seen and solved.

Geostatistics became practical with software. Many of the problems in this book are small neat problems that can be solved without the need for specialized software; however, in practice, specialized software will be required. There are a number of public domain and commercial software for geostatistics. Each software reflects the experience and preferences of the software creator/vendor and has specific strengths and weaknesses. It was beyond the scope of this book to review available software. In practice, it is a challenge to learn how to solve your geostatistical problem with the available software.

Model checking and validation are of unquestioned importance. Good estimates are ones that are close to the truth. Good distributions of uncertainty are ones that are narrow, yet fair. That is, the true value should come within our stated P_{10} and P_{90} about 80% of the time. We can check estimation and uncertainty with cross validation or new data. Models should reproduce the data and our deemed relevant statistical parameters. Problems that reveal best practice in model checking are warranted.

The challenge of spatial prediction in the presence of sparse data is important and will not go away. In fact, a greater variety of data and difficult decisions will be encountered in the future. Geological heterogeneity and the consequent uncertainty are important to quantify.

BIBLIOGRAPHY

Alabert, F.G. (1987). The practice of fast conditional simulations through the LU decomposition of the covariance matrix. *Mathematical Geology*, **19(5)**: 369–386.

Almeida, A.S. (1993), *Joint Simulation of Multiple Variables with a Markov-Type Coregionalization Model*, Ph.D. Thesis, Stanford University.

Almeida, A.S. and A.G. Journel (1994). Joint simulation of multiple variables with a markov-type coregionalization model, *Mathematical Geology*, **26**, 565–588.

Armstrong, M. (1984). Improving the estimation and modeling of the variogram, In *Geostatistics for Natural Resource Characterization*, G. Verly, M. David, A.G. Journel, and A. Marechal, eds. Reidel, Dordrecht, Holland, 1–19.

Armstrong, M. (1998). *Basic linear geostatistics*. Springer, Berlin, 153.

Bellman, R.E. (1961). *Adaptive control processes: a guided tour*. Princeton, New Jersey, 255.

Bogaert, P. (1999). On the optimal estimation of the cumulative distribution function in presence of spatial dependence. *Mathematical Geology*, **31(2)**: 213–239.

Bourgault, G. (1997). Spatial declustering weights. *Mathematical Geology*, **29(2)**: 277–290.

Caers J. (2001). Geostatistical Reservoir Modeling Using Statistical Pattern Recognition. *Journal of Petroleum Science and Engineering*, **29(3)**:177–188.

Chilès, J.P. and P. Delfiner (1999). *Geostatistics: modeling spatial uncertainty*. John Wiley & Sons Inc., New York, 695.

Christakos G. (1984). On the problem of permissible covariance and variogram models. *Water Resources Research*, **20**, 251–265.

Christakos, G. (1992). *Random field models in earth sciences*, Academic Press, San Diego, CA.

Clark, I. and W.V. Harper (2000). *Practical Geostatistics 2000*. Geostokos (Ecosse) Limited, Scotland, 342.

Cressie, N. and D.M. Hawkins (1980). Robust estimation of the variogram. *Mathematical Geology*, **12(2)**: 115–125.

Solved Problems in Geostatistics. By O. Leuangthong, K.D. Khan, and C.V. Deutsch

Cressie, N. (1985). Fitting variogram models by weighted least squares. *Mathematical Geology*, **17(5)**: 563–586.

Cressie, N. (1990). The origins of kriging. *Mathematical Geology*, **22(3)**: 239–252.

Cressie, N. (1991). *Statistics for spatial data.* Wiley Interscience, New York, 900.

Daly, C. and G. Verly (1994). Geostatistics and data integration, In *Geostatistics for the Next Century*, R. Dimitrakopoulos ed. Kluwer, 94–107.

David, M., (1977). *Geostatistical ore reserve estimation.* Elsevier, Amsterdam, The Netherlands, 364.

Davis, B.M. and L.E. Borgman (1979). Some exact sampling distributions for variogram estimators. *Mathematical Geology*, **11(6)**: 643–653.

Davis, J.C. (1986). *Statistics and data analysis in geology*, John Wiley & Sons Inc., New York.

Davis, M.W. (1987). Production of conditional simulations via the LU triangular decomposition of the covariance matrix. *Mathematical Geology*, **19(2)**: 91–98.

Deutsch, C.V. (1987). *A Probabilistic Approach to Estimate Effective Absolute Permeability.* M.Sc. Thesis, Stanford University, 165.

Deutsch, C.V. (1994). Kriging with strings of data. *Mathematical Geology*, **26(5)**: 623–637.

Deutsch, C.V. (1995). Correcting for negative weights in ordinary kriging. *Computers & Geosciences*, **22(7)**: 765–773.

Deutsch, C.V. and A.G. Journel (1998). *GSLIB: Geostatistical Software Library and Users Guide, 2nd Edition.* Oxford University Press, New York, 376.

Deutsch, C.V. (2002). *Geostatistical Reservoir Modeling.* Oxford University Press, New York, 376.

Deutsch, C.V., 2004. A Statistical resampling program for correlated data: spatial_bootstrap, *Centre for Computational Geostatistics Annual Report 6.* University of Alberta, Edmonton, Alberta, 9.

Doyen, P.M. (1988). Porosity from seismic data: A geostatistical approach. *Geophysics*, **53(10)**: 1263–1275.

Draper, N. and H. Smith (1966). *Applied regression analysis*, John Wiley & Sons Inc., New York.

Dubrule, O. (1983). Two methods with different objectives: splines and kriging, *Mathematical Geology*, **15(2)**: 245–257.

Efron, B. (1982). *The jackknife, the bootstrap and other resampling plans.* Society for Industrial and Applied Math, Philadelphia, PA, 92.

Efron, B. and R.J. Tibshirani (1993). *An introduction to the bootstrap*. Chapman & Hall, New York, 436.

Emery, X. and J. M. Ortiz (2005). Histogram and variogram inference in the multigaussian model, stochastic. *Environmental Research and Risk Assessment* (ISI), **19(1)**: 48–58.

Gandin, L. (1963). *Objective analysis of meteorological fields*, Gidrometeoizdat, Leningrad, 287. Translation (1965). Israel Program for Scientific Translations, Jerusalem.

Gendzwill, D.J. and M.R Stauffer (1981). Analysis of triaxial ellipsoids: their shapes, plane sections, and plane projections, *Mathematical Geology*, **13(2)**: 135–152.

Genton M.G. (1998a). Highly robust variogram estimation. *Mathematical Geology*, **30(2)**: 213–221.

Genton M.G. (1998b). Variogram fitting by generalized least squares using an explicit formula for the covariance structure. *Mathematical Geology*, **30(4)**: 323–345.

Glacken, I. M. (1996). Change of support and use of economic parameters for block selection. *Proceedings Fifth International Geostatistics Congress*; 811–821.

Goldberger, A. (1962). Best linear unbiased prediction in the generalized linear regression model, *JASA*, **57**: 369–375.

Gomez-Hernandez, J.J. and A.G. Journel (1992). Joint sequential simulation of multigaussian fields. *Geostatistics Troia 1992*, Vol. 1, Kluwer, 85–94.

Gomez-Hernandez, J.J. and X.H. Wen (1998). To be or not to be multi-gaussian? A reflection on stochastic hydrology. *Advances in Water Resources*. **21(1)**: 47–61.

Goovaerts, P. (1994). On a controversial method for modeling coregionalization. *Mathematical Geology*, **26**: 197–204.

Goovaerts, P. (1997). *Geostatistics for Natural Resources Evaluation*. Oxford University Press, New York, 483.

Goulard, M. (1989). Inference in a coregionalization model, *Geostatistics*, Vol. 1, Kluwer, 397–408.

Goulard, M. and M. Voltz (1992). Linear coregionalization model: tools for estimation and choice of cross-variogram matrix. *Mathematical Geology,* **24**: 269–286.

Gringarten, E., P. Frykman, and C.V. Deutsch (2000). Determination of reliable histogram and variogram parameters for geostatistical modeling. AAPG Hedberg Symposium, *Applied Reservoir Characterization using Geostatistics*, The Woodlands, Texas.

Gringarten, E. and C.V. Deutsch (2000). Teacher's aide: variogram interpretation and modeling. *Mathematical Geology*, **33(4)**: 507–534.

Guardiano, F. and R.M. Srivastava (1992). Multivariate geostatistics: Beyond bivariate moments. *Geostatistics Troia 1992*, Vol. 1, Kluwer, 133–144.

Hammersley, J.M. and D.C. Handscomb (1964). *Monte carlo methods*, John Wiley & Sons Inc., New York.

Haldorsen, H.H. and E. Damsleth (1990). Stochastic modeling. *Journal of Petroleum Technology*, Society of Petroleum Engineering, SPE 20321, 404–412.

Hall P. (1988). *Introduction to the theory of coverage processes*. John Wiley & Sons Inc., New York, 428.

Hoef, J.M.V. and N. Cressie (1993). Multivariable spatial prediction. *Mathematical Geology,* **25**: 219–240.

Hohn, M.E. (1988). *Geostatistics and Petroleum Geology*, Van Nostrand, New York.

Huijbregts, C.J. and G. Matheron (1971). Universal kriging – an optimal approach to trend surface analysis. In *Decision Making in the Mineral Industry*, Canadian Institute of Mining and Metallurgy, Special Vol. 12, 159–169.

Isaaks, E.H. and R.M. Srivastava (1989). *An introduction to applied geostatistics*. Oxford, New York, 561.

Isaaks, E.H. (1990). *The application of Monte Carlo methods to the analysis of spatially correlated data*. PhD Thesis, Stanford University, Stanford, CA.

Johnson, M. (1987). *Multivariate statistical simulation*. John Wiley & Sons Inc., New York.

Johnson, N.L. and S. Kotz (1970). *Continuous univariate distributions – 2*. John Wiley & Sons Inc., New York., 752.

Johnson, R.A. and D.W. Wichern (1998). *Applied multivariate statistical analysis*, 4th Edition. Prentice-Hall, New Jersey, 816.

Journel A.G. (1974). Geostatistics for conditional simulation of ore bodies. *Econom. Geology*, Vol. 69, 673–687.

Journel A.G. (1982). The indicator approach to estimation of spatial distributions. *Proceedings of the 17th International APCOM Symposium*, Society of Mining Engineers, 793–806.

Journel A.G. (1983). Nonparameteric estimation of spatial distributions. *Mathematical Geology*, **15(3)**: 445–468.

Journel A.G. (1984). The place of non-parametric geostatistics. *Geostatistics for Natural Resources Characterization*, Vol. 1, Reidel, Dordrecht, Holland, 307–355.

Journel, A.G. (1989). *Fundamentals of Geostatistics in Five Lessons*. American Geophysical Union, Washington, 40.

Journel A.G. (1994). Modeling uncertainty: Some conceptual thoughts. *Geostatistics for the Next Century*, Kluwer, 30–43.

Journel, A.G. (1999a). Conditioning geostatistical operations to nonlinear volume averages. *Mathematical Geology*, **31(8)**: 931–953.

Journel, A.G. (1999b). Markov models for cross-covariances. *Mathematical Geology*, **31(8)**, 955–964.

Journel, A.G. (2002). Combining knowledge from diverse sources: An alternative to traditional data independence hypotheses. *Mathematical Geology*, **34(5)**, 573–596.

Journel A.G and C.V. Deutsch (1993). Entropy and Spatial Disorder. *Mathematical Geology*. **25(3)**:329–355.

Journel, A.G. and C.J. Huijbregts (1978). *Mining geostatistics*. Academic Press, London, 600.

Journel A.G and D. Posa (1990). Characteristic behavior and order relations for indicator variograms. *Mathematical Geology*. **22(8)**:1011–1025.

Journel, A.G. and M.E. Rossi (1989). When do we need a trend model in kriging? *Mathematical Geology*, **21(7)**: 715–739.

Krige, D.G. (1951). *A Statistical Approach to Some Mine Valuations and Allied Problems at the Witwatersrand*. Master's thesis, University of Witwatersrand.

Kupfersberger, H., C.V. Deutsch, and A.G. Journel (1998). Deriving constraints on small-scale variograms due to variograms of large-scale data. *Mathematical Geology*, **30(7)**: 837–852.

Mallet, J.L. (2002). *Geomodeling*, Oxford University Press, New York, 599.

Mantoglou, A. and J.L. Wilson (1982). The turning bands method for simulation of random fields using line generation by a spectral method. *Water Resources Research*, **18(5)**: 1379–1394.

Marechal, A. (1984). Kriging seismic data in presence of faults, in, *Geostatistics for Natural Resource Characterization*, G. Verly, M. David, A.G. Journel and A. Marechal, eds. Reidel, Dordrecht, Holland, Part 1, 385–420.

Matern, B. (1960). Spatial variation. *Lecture Notes in Statistics*, Vol. **36**, Springer-Verlag, New York, 151.

Matheron, G. (1969). *Le Krigeage Universel: Fascicule 1.* Cahiers du CMM, 82.

Matheron, G. (1971). *The theory of regionalized variables and its applications.* In Cahiers du CMM. No. 5, Ecole nationale superieure des mines de Paris, Fontainebleau, France, 211.

Matheron, G. (1973). The intrinsic random functions and their applications. *Advances Applied Probability,* **5**, 439–468.

Matheron, G. (1975). *Random Sets and Integral Geometry.* John Wiley & Sons Inc., New York, 261.

Myers, D. E. (1982). Matrix formulation of co-kriging. *Mathematical Geology*, **14(3)**: 249–257.

Myers, D. E. (1991). Pseudo-cross variograms, positive-definiteness, and cokriging. *Mathematical Geology*, **23(6)**: 805–816.

Olea, R.A. (1991). *Geostatistical glossary and multilingual dictionary.* International Association for Mathematical Geology, Monograph 3. Oxford, New York, 177.

Olea, R.A. (2007). Declustering of clustered preferential sampling for histogram and semivariogram inference. *Mathematical Geology*, **39(5)**: 453–467.

Omre, H. (1984). The variogram and its estimation. In *Geostatistics for Natural Resource Characterization*, G. Verly, , M. David., A.G. Journel, and A. Marechal, eds. Reidel, Dordrecht, Holland, 107–125.

Ortiz, J.M. (2003). *Characterization of high order correlation for enhanced indicator simulation*. PhD Thesis, University of Alberta, 287.

Oz, B., C.V. Deutsch and P. Frykman (2002). A visual basic program for histogram and variogram scaling. *Computers & Geosciences*, **28(1)**: 21–31.

Parker, H.M. (1980). The volume-variance relationship: a useful tool for mine planning. In *Geostatistics*, P. Mousset-Jones, eds. McGraw Hill, New York, 61–91.

Pyrcz, M.P. and C.V. Deutsch (2003). Declustering and debiasing. Downloaded in October 2007 from *http://www.gaa.org.au/pdf/DeclusterDebias-CCG.pdf.*

Reklaitis, G.V., A. Ravindran, and K.M. Ragsdell (1983). *Engineering Optimization*, John Wiley & Sons Inc., New York, 684.

Rivoirard, J. (2001). Which models for collocated cokriging? *Mathematical Geology*, **33(2)**: 117–131.

Rossi, M.E. and D. Posa (1992). Measuring departure from Gaussian assumptions in spatial processes. *Proceedings of the 23rd International APCOM Symposium*, 189–197.

Schofield, N. (1993). Using the entropy statistics to infer population parameters from spatially clustered sampling. In *Geostatistics Troia '92*, A. Soares, ed. Vol. 1, Kluwer Academic, Dordrecht, Holland, 109–119.

Shmaryan, L.E. and A.G. Journel (1999). Two markov models and their application. *Mathematical Geology*, **31(8)**: 965–988.

Sichel, H.S. (1952). New methods in the statistical evaluation of mine sampling data. *Transactions of the Institution for Mining and Metallurgy*, London, 261–288.

Spiegel, M.R. (1980). *Theory and problems of probability and statistics.* Schaum's Outline Series, McGraw Hill, New York, 372.

Srivastava, R.M. (1985). Some notes on the second-order moments of the 3D bombing model, Internal Report, Department of Applied Earth Sciences, Stanford University.

Srivastava, R. M. (1987). Minimum variance or maximum profitability? *CIM Bulletin*, **80(901)**: 63–68.

Stein M.L. (2002). The screening effect in kriging. *The Annals of Statistics*, **30(1)**: 298–323.

Strebelle, S. and A. G. Journel (2001). Reservoir modeling using multiple-point statistics. *2001 SPE Annual Technical Conference and Exhibition,* Society of Petroleum Engineers, New Orleans, LA, SPE 71324.

Strebelle, S. (2002). Conditional simulation of complex geological structures using multiple-point statistics: *Mathematical Geology*, **34**, 1–22.

van Lieshout M.N.M and E.W. van Zwet (2000). Maximum likelihood estimation for the bombing model. *Centrum voor Wiskunde en Informatica*, Amsterdam, PNA-R0008.

Vargas-Guzman, J.A., A.W. Warrick, and D.E. Myers (1999). Multivariate correlation in the framework of support and spatial scales of variability. *Mathematical Geology,* **31(1)**: 85–103.

Verly, G. (1983). The multigaussian approach and its applications to the estimation of local reserves. *Mathematical Geology*, **15(2)**: 259–286.

Verly, G. W. (1984). The block distribution given a point multivariate normal distribution. *Geostatistics for Natural Resources Characterization*. Reidel, Dordrecht, Holland, 495–514.

Wackernagel, H., P. Petitgas and Y. Touffait (1989). Overview of methods for coregionalization analysis. *Geostatistics*, Vol. 1, Kluwer, 409–420.

Wackernagel H. (1994). Cokriging versus kriging in regionalized multivariate data analysis. *Geoderma*, **62**: 83–92.

Wackernagel H. (2003). *Multivariate Geostatistics*. 3rd edition, Springer, 387.

Xu, W., T.T. Tran, R.M. Srivastava and A.G. Journel (1992). Integrating seismic data in reservoir modeling: the collocated cokriging alternative, *67th Annual Technical Conference and Exhibition,* Society of Petroleum Engineers, Washington, DC, SPE 24742.

Yarus, J.M. (1995). Selected readings on geostatistics, stochastic modeling and geostatistics: principles, methods, and case studies, *AAPG Computer Applications in Geology*, **3**, 369–370.

Zhang, T., P. Switzer, and A. G. Journel (2006). Filter-based classification of training image patterns for spatial simulation. *Mathematical Geology*, **38**, 63–80.

Zinn, B. and C.F. Harvey (2003). When good statistical models of aquifer heterogeneity go bad: A comparison of flow, dispersion, and mass transfer in connected and multivariate Gaussian hydraulic conductivity fields. *Water Resources Research*, **39(3)**, 19p.

INDEX

Solved Problems in Geostatistics. By O. Leuangthong, K.D. Khan, and C.V. Deutsch